Extended Finite Element Method

Extended Finite Element Method

Zhuo Zhuang, Zhanli Liu, Binbin Cheng,
and Jianhui Liao

Tsinghua University, Beijing

ELSEVIER

AMSTERDAM • BOSTON • HEIDELBERG • LONDON
NEW YORK • OXFORD • PARIS • SAN DIEGO
SAN FRANCISCO • SINGAPORE • SYDNEY • TOKYO

Academic Press is an Imprint of Elsevier

Academic Press is an imprint of Elsevier
The Boulevard, Langford Lane, Kidlington, Oxford OX5 1GB, UK
225 Wyman Street, Waltham, MA 02451, USA

First edition 2014

British Library Cataloguing in Publication Data
A catalogue record for this book is available from the British Library

Library of Congress Cataloging-in-Publication Data
A catalog record for this book is availabe from the Library of Congress

ISBN: 978-0-12-407717-1

For information on all Academic Press publications
visit our web site at store.elsevier.com

Printed and bound in the US

14 15 16 17 18   10 9 8 7 6 5 4 3 2 1

Working together
to grow libraries in
developing countries

www.elsevier.com • www.bookaid.org

Contents

Preface   ix

1. **Overview of Extended Finite Element Method**   **1**
    1.1   **Significance of Studying Computational Fracture Mechanics**   **1**
    1.2   **Introduction to X-FEM**   **2**
    1.3   **Research Status and Development of X-FEM**   **8**
          1.3.1   The Development of X-FEM Theory   8
          1.3.2   Development of 3D X-FEM   10
    1.4   **Organization of this Book**   **11**

2. **Fundamental Linear Elastic Fracture Mechanics**   **13**
    2.1   **Introduction**   **13**
    2.2   **Two-Dimensional Linear Elastic Fracture Mechanics**   **15**
    2.3   **Material Fracture Toughness**   **19**
    2.4   **Fracture Criterion of Linear Elastic Material**   **20**
    2.5   **Complex Fracture Criterion**   **22**
          2.5.1   Maximum Circumference Tension Stress Intensity
                  Factor Theory   22
          2.5.2   Minimum Strain Energy Density Stress Intensity Fac-
                  tor Theory   24
          2.5.3   Maximum Energy Release Rate Theory   27
    2.6   **Interaction Integral**   **29**
    2.7   **Summary**   **31**

3. **Dynamic Crack Propagation**   **33**
    3.1   **Introduction to Dynamic Fracture Mechanics**   **33**
    3.2   **Linear Elastic Dynamic Fracture Theory**   **35**
          3.2.1   Dynamic Stress Field at Crack Tip Position   35
          3.2.2   Dynamic Stress Intensity Factor   37
          3.2.3   Dynamic Crack Propagating Condition and Velocity   38
    3.3   **Crack Driving Force Computation**   **41**
          3.3.1   Solution Based on Nodal Force Release   41
          3.3.2   Solution Based on Energy Balance   43
    3.4   **Crack Propagation in Steady State**   **44**
    3.5   **Engineering Applications of Dynamic Fracture Mechanics**   **45**
    3.6   **Summary**   **49**

4. **Fundamental Concept and Formula of X-FEM**   **51**
   4.1   X-FEM Based on the Partition of Unity   **51**
   4.2   Level Set Method   **53**
   4.3   Enriched Shape Function   **55**
         4.3.1   Description of a Strong Discontinuity Surface   55
         4.3.2   Description of a Weak Discontinuity Surface   58
   4.4   Governing Equation and Weak Form   **60**
   4.5   Integration on Spatial Discontinuity Field   **64**
   4.6   Time Integration and Lumped Mass Matrix   **67**
   4.7   Postprocessing Demonstration   **68**
   4.8   One-Dimensional X-FEM   **68**
         4.8.1   Enriched Displacement   68
         4.8.2   Mass Matrix   72
   4.9   Summary   **72**

5. **Numerical Study of Two-Dimensional Fracture
   Problems with X-FEM**   **75**
   5.1   Numerical Study and Precision Analysis of X-FEM   **76**
         5.1.1   A Half Static Crack in a Finite Plate   76
         5.1.2   A Beam with Stationary Crack under Dynamic
                 Loading   77
         5.1.3   Simulation of Complex Crack Propagation   77
         5.1.4   Simulation of the Interface   79
         5.1.5   Interaction Between Crack and Holes   81
         5.1.6   Interfacial Crack Growth in Bimaterials   83
   5.2   Two-Dimensional High-Order X-FEM   **84**
         5.2.1   Spectral Element-Based X-FEM   84
         5.2.2   Mixed-Mode Static Crack   87
         5.2.3   Kalthoff's Experiment   89
         5.2.4   Mode I Moving Crack   91
   5.3   Crack Branching Simulation   **93**
         5.3.1   Crack Branching Enrichment   94
         5.3.2   Branch Criteria   95
         5.3.3   Numerical Examples   97
   5.4   Summary   **101**

6. **X-FEM on Continuum-Based Shell Elements**   **103**
   6.1   Introduction   **104**
   6.2   Overview of Plate and Shell Fracture Mechanics   **104**
         6.2.1   Kirchhoff Plate and Shell Bending Fracture Theory   105
         6.2.2   Reissner Plate and Shell Bending Fracture Theory   109
   6.3   Plate and Shell Theory Applied In Finite Element Analysis   **113**
   6.4   Brief Introduction to General Shell Elements   **115**
         6.4.1   Belytschko–Lin–Tsay Shell Element   115
         6.4.2   Continuum-Based Shell Element   116
   6.5   X-FEM on CB Shell Elements   **119**

	6.5.1	Shape Function of a Crack Perpendicular to the Mid-Surface	119
	6.5.2	Shape Function of a Crack Not Perpendicular to the Mid-Surface	123
	6.5.3	Total Lagrangian Formulation	125
	6.5.4	Time Integration Scheme and Linearization	127
	6.5.5	Continuum Element Transformed to Shell	128
6.6	**Crack Propagation Criterion**		**129**
	6.6.1	Stress Intensity Factor Computation	129
	6.6.2	Maximum Energy Release Rate Criterion	133
6.7	**Numerical Examples**		**136**
	6.7.1	Mode I Central Through-Crack in a Finite Plate	136
	6.7.2	Mode III Crack Growth in a Plate	137
	6.7.3	Steady Crack in a Bending Pipe	137
	6.7.4	Crack Propagation Along a Given Path in a Pipe	139
	6.7.5	Arbitrary Crack Growth in a Pipe	140
6.8	**Summary**		**140**

7.　Subinterfacial Crack Growth in Bimaterials　143

7.1	**Introduction**		**143**
7.2	**Theoretical Solutions of Subinterfacial Fracture**		**144**
	7.2.1	Complex Variable Function Solution for Sub-interfacial Cracks	144
	7.2.2	Solution Considering the Crack Surface Affected Area	147
	7.2.3	Analytical Solution of a Finite Dimension Structure	149
7.3	**Simulation of Subinterfacial Cracks Based On X-FEM**		**153**
	7.3.1	Experiments on Subinterfacial Crack Growth	153
	7.3.2	X-FEM Simulation of Subinterfacial Crack Growth	155
7.4	**Equilibrium State of Subinterfacial Mode I Cracks**		**158**
	7.4.1	Effect on Fracture Mixed Level by Crack Initial Position	158
	7.4.2	Effect on Material Inhomogeneity and Load Asymmetry	159
7.5	**Effect on Subinterfacial Crack Growth from a Tilted Interface**		**163**
7.6	**Summary**		**165**

8.　X-FEM Modeling of Polymer Matrix Particulate/ Fibrous Composites　167

8.1	**Introduction**		**167**
8.2	**Level Set Method for Composite Materials**		**169**
	8.2.1	Level Set Representation	169
	8.2.2	Enrichment Function	172
	8.2.3	Lumped Mass Matrix	173
8.3	**Microstructure Generation**		**175**
8.4	**Material Constitutive Model**		**176**

8.5	Numerical Examples	**177**
	8.5.1 Static Analysis	177
	8.5.2 Dynamic Analysis	181
8.6	Summary	**187**

9. **X-FEM Simulation of Two-Phase Flows** — **189**

9.1	Governing Equations and Interfacial Conditions	**189**
9.2	Interfacial Description of Two-Phase Flows	**192**
9.3	X-FEM and Unknown Parameters Discretization	**194**
9.4	Discretization of Governing Equations	**200**
9.5	Numerical Integral Method	**205**
9.6	Examples and Analyses	**207**
9.7	Summary	**211**

10. **Research Progress and Challenges of X-FEM** — **213**

10.1	Research on Micro-Scale Crystal Plasticity	**213**
	10.1.1 Discrete Dislocation Plasticity Modeling	215
	10.1.2 X-FEM Simulation of Dislocations	219
10.2	Application of Multi-Scale Simulation	**223**
10.3	Modeling of Deformation Localization	**224**
10.4	Summary	**228**

Appendix A: Westergaard Stress Function Method	229
Appendix B: *J* Integration	245
References	259
Index	269

The extended finite element method (X-FEM) is a novel numerical methodology, which was first proposed by Belytschko et al. in 1999. It has subsequently been developed very quickly in the mechanics field worldwide. Based on the finite element method and fracture mechanics theory, X-FEM can be applied to solve complicated discontinuity issues including fracture, interface, and damage problems with great potential for use in multi-scale computation and multi-phase coupling problems.

The fundamental concept and formula of X-FEM are introduced in this book, as well as the technical process of program implementation. The expressions for enriched shape functions of the elements are provided, which include the displacement discontinued crack, termed as "strong discontinuity", and strain discontinued interface, termed as "weak discontinuity", like heterogeneous materials with voids and inclusions, interfaces of bimaterial and two-phase flows. X-FEM can be used to simulate element-crossed cracks and element-embedded cracks. Cracks with complex geometry can be modeled by structured meshes and can propagate along arbitrary route in the elements without the need for a re-meshing process, which provides considerable savings in computation cost whilst achieving precision.

In the early 1990s, the first author of this book, Prof. Zhuo Zhuang, was working on Ph.D. research under Prof. Patraic O'Donoghue at University College Dublin, Ireland, and completed a thesis on the development of the finite element method for dynamic crack propagation in gas pipelines. In 1995, he returned to China and took an academic position at Tsinghua University. He has the privilege of learning from and working with Prof. Keh-Chih Hwang, and is striving to simulate arbitrary crack growth in three-dimensional continuities and curved shells. This is a natural choice for crack propagation, in which the original failure behavior of the structures reappears. In 2011, this aspiration was released by Dr. Binbin Cheng, who is the third author of this book. From his Ph.D. thesis work at Tsinghua University, they have developed an X-FEM code, named SAFRAC, with its own properties. The second author, Dr. Zhanli Liu, obtained a doctoral degree in 2009. The research thesis "The Investigation of Crystal Plasticity at Microscale by Discrete Dislocation and Nonlocal Theory" was nominated and achieved a national excellent doctor degree thesis in 2011. After graduation from Tsinghua University, he went to Northwestern University, USA to conduct postdoctoral research under Prof. Ted Belytschko. He continued to develop the X-FEM method for dynamic crack propagation and applications in heterogeneous materials, like ultrasonic wave propagation in three-dimensional polymer matrices enhanced by particles and short fibers. He returned to Tsinghua

University and took an academic position in 2012. The fourth author, Dr. Jianhui Liao, obtained a doctoral degree in 2011 at Tsinghua University. His thesis work focuses on the application of X-FEM simulation in two-phase flows. One professor and three former Ph.D. students are working together to write this book in order to demonstrate the research achievements on X-FEM in the last decade.

In this book, Chapter 1 reviews the development history, reference summarization, and research actuality of X-FEM. Chapters 2 and 3 provide an introduction to fracture mechanics, considering the essential concepts of static and dynamic linear elastic fracture mechanics, such as the crack propagation criterion, the calculation of stress intensity factor by interaction integral, the nodal force release technique to simulate crack propagation in conventional FEM, and so on. These two chapters provide essential knowledge of fracture mechanics essential for study of the subsequent chapters. Readers who are familiar with fracture mechanics can skip these two chapters. Chapters 4 and 5 contain the basic ideas and formulations of X-FEM. Chapter 4 focuses on the theoretical foundation, mathematical description of the enrichment shape function, discrete formulation, etc. In Chapter 5, based on the program developed by the authors and their co-workers, numerical studies of two-dimensional fracture problems are provided to demonstrate the capability and efficiency of the algorithm and the X-FEM program in applications of strong and weak discontinuity problems. In Chapters 6—9, scientific research conducted by the author's group is given as examples to introduce the applications of X-FEM. In Chapter 6, a novel theory formula and computational method of X-FEM is developed for three-dimensional (3D) continuum-based (CB) shell elements to simulate arbitrary crack growth in shells using the concept of enriched shape functions. In Chapter 7, the algorithm is discussed and a program is developed based on X-FEM for simulating subinterfacial crack growth in bimaterials. Numerical analyses of the crack growth in bimaterials provide a clear description of the effect on fracture of the interface and loading. In Chapter 8, a method for representing discontinuous material properties in a heterogeneous domain by X-FEM is applied to study ultrasonic wave propagation in polymer matrix particulate/fibrous composites. In Chapter 9, a simulation method of transient immiscible and incompressible two-phase flows is proposed, which demonstrates how to deal with multi-phase flow problems by applying X-FEM methodology. Based on the scientific research in the author's group, Chapter 10 gives the applications of X-FEM in other frontiers of mechanics, e.g. nano-mechanics, multi-scale simulations, crack branches, and so on.

This book was published in a Chinese version in 2012, and was the first book on X-FEM published in China. At that time, Dr. Zhanli Liu was a postdoctoral fellow working at Northwestern University in the USA. He presented a copy of the book to Prof. Ted Belytschko to express our respect for him. Ted was very happy to see it and made complimentary remarks about the book, although he could not follow the Chinese characters but only the equations and figures. He encouraged us to publish this book in an English version.

Regarding the English version, we would like to thank Mr. Lei Shi, Ms. Qiuling Zhang, and Ms. Hongmian Zhao at Tsinghua University Press. Without their encouragement and help, we could not have completed this book. We are also grateful to the Ph.D. candidates Ms. Dandan Xu and Mr. Qinglei Zeng for the computational examples that they provided.

This book is suitable for teachers, engineers, and graduate students on the disciplines of mechanics, civil engineering, mechanical engineering, and aerospace engineering. It can also be referenced by X-FEM program developers.

Zhuo Zhuang
Zhanli Liu
Binbin Cheng
Jianhui Liao
October 2013

Chapter 1

Overview of Extended Finite Element Method

Chapter Outline
1.1 Significance of Studying
Computational Fracture
Mechanics   1
1.2 Introduction to X-FEM   2
1.3 Research Status and
Development of X-FEM   8

1.3.1 The Development
of X-FEM Theory   8
1.3.2 Development of 3D
X-FEM   10
1.4 Organization of this Book   11

1.1 SIGNIFICANCE OF STUDYING COMPUTATIONAL FRACTURE MECHANICS

Fracture is one of the most important failure modes. In various engineering fields, many catastrophic accidents have started from cracks or ends at crack propagation, such as the cracking of geologic structures and the collapse of engineering structures during earthquakes, damage of traffic vehicles during collisions, the instability crack propagation of pressure pipes, and the fracture of mechanical components. These accidents have caused great loss to people's lives and economic property. However, usually it is very difficult to quantitatively provide the causes of crack initiation. So research on fracture mechanics, which is mainly focused on studying the propagation or arrest of initiated cracks, is of great theoretical importance and has broad application potential.

Modern fracture mechanics has been booming and has been studied extensively in recent years; this is because it is already deeply rooted in the modern high-technology field and engineering applications. For example, large-scale computers facilitate the numerical simulation of complicated fracture processes, and new experimental techniques provided by modern physics, such as advanced scanning electron microscope (SEM) analysis, surface analysis, and high-speed photography, make it possible to study the fracture process from the micro-scale to the macro-scale. This understanding

Extended Finite Element Method. http://dx.doi.org/10.1016/B978-0-12-407717-1.00001-7

of the basic laws of fracture plays an important role in theoretically guiding the applications of fracture mechanics in engineering, such as the toughening of new materials, the development of biological and biomimetic materials, the seismic design of buildings and nuclear reactors, the reliability of microelectronic components, earthquake prediction in geomechanics, the exploration and storage of oil and gas, the new design of aerospace vehicles, etc. After integration with modern science and high-technology methods, fracture mechanics is taking on a new look.

Cracks in reality are usually in three dimensions, and have complicated geometries and arbitrary propagation paths. For a long time, one of the difficult challenges of mechanics has been to study crack propagation along curved or kinked paths in three-dimensional structures. In these situations, the "straight crack" assumption in conventional fracture mechanics is no longer valid, so theoretical methods are very limited for this problem. Experiments are another important way to study the propagation of curved cracks, but most results are empirical and phenomenological, and mainly focus on planar cracks. In recent decades, numerical simulations have developed rapidly along with the development of computer technology. The new progress in computational mechanics methods, such as the finite element method, boundary element method, etc., provides the possibility of solving the propagation of curved cracks. Modeling crack propagation in three-dimensional solids and curved surfaces has become one of the hottest topics in computational mechanics. Computational fracture mechanics methods roughly include the finite element method with adaptive mesh (Miehe and Gürses, 2007), nodal force release method (Zhuang and O'Donoghue, 2000a, b), element cohesive model (Xu and Needleman, 1994), and embedded discontinuity model (Belytschko et al., 1988). All of these methods have some limitations when dealing with cracks with complicated geometries, such as when the crack path needs to be predefined, the crack must propagate along the element boundary, the computational cost is high, etc. In the last decades, the extended finite element method (X-FEM) proposed in the late 1990s has become one of the most efficient tools for numerical solution of complicated fracture problems.

1.2 INTRODUCTION TO X-FEM

One of the greatest contributions the scientists made to mankind in the twentieth century was the invention of the computer, which has greatly promoted the development of related industry and scientific research. Taking computational mechanics as an example, many new methods, such as the finite element method, finite difference, and finite volume methods have rapidly developed as the invention of computer. Thanks to these methods, a lot of traditional problems in mechanics can be simulated and analyzed numerically; more importantly, a number of engineering and scientific problems can be modeled and solved. As the development of modern information technology and computational science continues, simulation-based engineering and science has become helpful to scientists in exploring the mysteries of science, and provides an effective tool for the engineer to implement engineering innovations or product development with high reliability. The finite element method (FEM) is just one of the powerful tools of simulation-based engineering and science.

Since the appearance of the first FEM paper in the mid-1950s, many papers and books on this issue have been published. Some successful experimental reports and a series of articles have made great contributions to the development of FEM. From the 1960s, with the emergence of finite element software and its rapid applications, FEM has had a huge impact on computer-aided engineering analysis. The appearance of numerous advanced software not only meets the requirement of simulation-based engineering and science, but also promotes further development of the finite element method itself. If we compare a finite element to a large tree, it is like the growth of several important branches, like hybrid elements, boundary elements, the meshless method, extended finite elements, etc., make this particular tree prosper.

In analysis by the conventional finite element method, the physical model to be solved is divided into a series of elements connected in a certain arrangement, usually called the "mesh". However, when there are some internal defects, like interfaces, cracks, voids, inclusions, etc. in the domain, it will create some difficulties in the meshing process. On one hand, the element boundary must coincide with the geometric edge of the defects, which will induce some distortion in the element; on the other hand, the mesh size will be dependent on the geometric size of the small defects, leading to a nonuniform mesh distribution in which the meshes around the defects are dense, while those far from defects are sparse. As we know, the smallest mesh size decides the critical stable time increment in explicit analysis. So the small elements around the defects will heavily increase the computational cost. Also, defects, like cracks, can only propagate along the element edge, and not flow along a natural arbitrary path. Aiming at solving these shortcomings by using the conventional FEM to solve crack or other defects with discontinuous interfaces, Belytschko and Moës proposed a new computational method called the "extended finite element method (X-FEM)" (Belytschko and Black, 1999; Moës et al., 1999), and made an important improvement to the foundation of conventional FEM. In the last 10 years, X-FEM has been constantly improved and developed, and has already become a powerful and promising method for dealing with complicated mechanics problems, like discontinuous field, localized deformation, fracture, and so on. It has been widely used in civil engineering, aviation and space, material science, etc.

The core idea of X-FEM is to use a discontinuous function as the basis of a shape function to capture the jump of field variables (e.g., displacement) in the computational domain. So in the calculations, the description for the discontinuous field is totally mesh-independent. It is this advantage that makes it very suitable for dealing with fracture problems. Figure 1.1 is an example of a three-dimensional fracture simulated by X-FEM (Areias and Belytschko, 2005b), in which we find that the crack surface and front are independent of the mesh. Figure 1.2 demonstrates the process of transition crack growth under impact loading on the left lower side of a plane plate, in which we can investigate impact wave propagation in the plate and the stress singularity field at the crack tip location. In addition, it is very convenient to model the crack with complex geometries using X-FEM; one example of crack branching simulation is given in Figure 1.3 (Xu et al., 2013).

X-FEM is not only used to simulate cracks, but also to simulate heterogeneous materials with voids and inclusions (Belytschko et al., 2003b; Sukumar et al., 2001). The main difference is that: for cracks, the discontinuous field at the crack surface is the displacement; for inclusions, the derivative of displacement with respect to a spatial coordinate — the strain — is discontinuous. These two situations are defined as strong discontinuity (jump of displacement field) and weak discontinuity (jump of derivative of displacement with respect to spatial coordinate) respectively. Two different enrichment shape functions will be used to capture the two different discontinuities. Figure 1.4 shows an example of studying the effective modulus of carbon nanotube composites by X-FEM modeling. In the simulations, the mesh boundary does not have to coincide with the material interfaces, so the representative volume element (RVE) can be meshed by brick elements, which will greatly increase the efficiency of modeling.

The other advantage of X-FEM is that it can make use of known analytical solutions to construct the shape function basis, so accurate results can be obtained even using a relatively coarse mesh. When applying conventional FEM to model the singular field, like the stress field near a crack tip or dislocation core, a very dense mesh has to be used. However, in X-FEM, by introducing the known displacement solution of cracks or dislocations into the enrichment shape function, a satisfied solution can be obtained under a relatively coarse mesh. Figure 1.5 shows a plate with an initial crack at the left

FIGURE 1.1 X-FEM modeling of 3D fracture: displacement is magnified 200 times (Areias and Belytschko, 2005b).

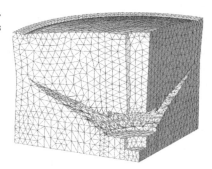

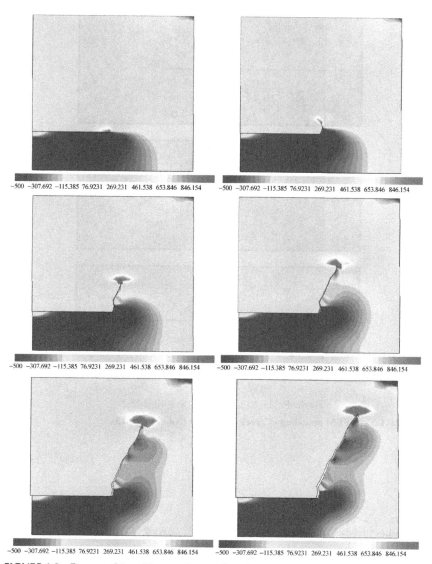

FIGURE 1.2   Process of transition crack growth under impact on the left lower side.

edge; the stress intensity factor can be calculated as a function of crack length. In the X-FEM simulation, without using the fine mesh near the crack tip, the calculated stress intensity factor (SIF) for 41 by 41 uniform elements can compare well with analytic solution.

It is worth pointing out that, other methods, like the boundary element method and meshless method, also have important applications on solving

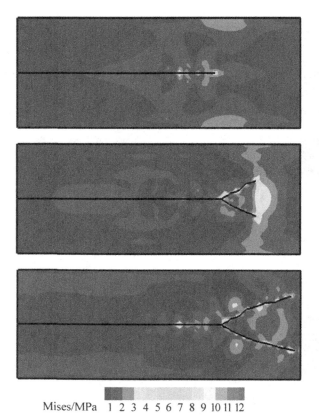

Mises/MPa 1 2 3 4 5 6 7 8 9 10 11 12

FIGURE 1.3   X-FEM modeling of crack branching (Xu et al., 2013).

FIGURE 1.4   X-FEM model of nanotube composites (Belytschko et al., 2003b).

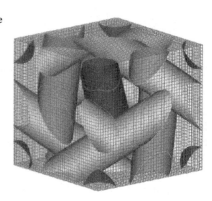

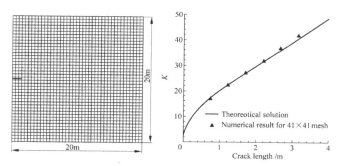

FIGURE 1.5   Stress intensity factor of a static crack in a finite plate.

discontinuous problems (Blandford et al., 1981; Belytschko et al., 1994). However, some inherent flaws limit their promotion: for example, the boundary element method is not good at dealing with problems with strong nonlinearity, heterogeneity, and so on; the meshless method lacks a solid theoretical foundation and rigorous mathematical proof, so there are still some uncertain parameters like the radius of the interpolation domain, background integration domain, etc.; commercial software still does not have mature modules for these two methods. In contrast, X-FEM is developed under the standard framework of FEM, and retains all the advantages of the conventional FEM method. Some commercial software, like ABAQUS and LS-DYNA, already have a basic X-FEM module for fracture simulations.

Given the above, the characterizations and advantages of X-FEM can be summarized as follows:

1. It allows for crack location and propagation inside the elements; cracks with complex geometry can be modeled by structured meshes and can propagate element by element without remeshing, which will greatly save computational cost.
2. The elements containing crack surfaces and crack tips are enriched with additional degrees of freedom, so that the discontinuous shape function is used to capture the singularity of the stress field near the crack tip. An accurate solution can therefore be obtained using a coarse mesh.
3. Compared with the remeshing technique in FEM, mapping of field variables after crack propagation is not necessary in X-FEM.
4. Compared with the boundary element method, X-FEM is applicable for multi-material or multi-phase problems, especially problems with geometric and contact nonlinearities.
5. It is convenient to implement in commercial software and with parallel computing.

All the features above illustrate why X-FEM has many successful applications.

1.3 RESEARCH STATUS AND DEVELOPMENT OF X-FEM

Since first being proposed in 1999, X-FEM has been growing and developing for more than 10 years. The main progress achieved can be roughly summarized as the following two aspects: one is the improvement of related theories, like those dealing with blending elements, the subdomain integration of a discontinuous field, stability of explicit time integration, lumped scheme of additional mass matrix, the extent of the basis of enrichment shape function, and so on (these topics will be described in detail in the following chapters); the other is the development of element types, from 2D plane element to 3D solid elements and shell elements. More and more element types with X-FEM features are being developed. Currently this area is still one of the main directions in the development of X-FEM.

1.3.1 The Development of X-FEM Theory

Belytschko and Black (1999) first proposed the idea of simulating crack propagation without changing the initial finite element mesh. Compared to the conventional FEM, the most significant innovation in this idea is that the element nodes near the crack surface or tip are enriched with the known displacement field to describe the appearance of the crack. Later, Moës et al. (1999) introduced the Heaviside function and crack tip function as the enrichment shape function of elements including the crack surface and tip respectively. They called this new method the eXtended Finite Element Method (X-FEM). Then, by using more than one enrichment shape function in crack tip elements, crack branching was successfully simulated by Daux et al. (2000). Later, a new crack nucleation criterion, the loss of hyperbolas, was introduced into X-FEM by Belytschko et al. (2003a). The crack propagation path and velocity can be well predicted using this criterion.

When the elements including crack surfaces and tips are enriched, for the neighboring elements, some nodes that are connected with enriched element have both standard and enriched degrees of freedom, whilst some nodes that are not connected with enriched elements only have standard degrees of freedom. Such elements are called "blending elements". The blending elements will decrease the efficiency of computation since they do not satisfy the partition of unity. Chessa et al. (2003) improved the performance of blending elements by using the assumed strain method. Legay et al. (2005) found that the convergence rate of blending elements increases with the increase of element order, and pointed out that special handling is not necessary for blending elements in high order elements. In early X-FEM, only the nodes nearest to the crack tip were enriched. Ventura et al. (2005) investigated the influence of enriched domain size around the crack tip on the convergence speed: they found that either increasing the domain of enrichment elements or refining the mesh in the fixed enriched domain can speed up the convergence.

In the X-FEM simulation of fracture, the displacement and stress field can be solved accurately. However, a postprocessing module is necessary to calculate the stress intensity factor (SIF) at the crack tip using the contour integral or least square method. To find an accurate SIF without using additional postprocessing, Liu et al. (2004) used not only the first term, but also the second term of the asymptotic displacement field near the crack tip, to enrich the element nodes near the crack tip. Their numerical results obtained with the reduced integration method show an improvement of the SIF calculation without postprocessing.

Song et al. (2006) used overlay element and phantom nodes to describe cracked elements by rearranging the basis function and nodal degree of freedom. This method is very suitable for the implementation of X-FEM in commercial software since no additional degrees of freedom are introduced in the cracked elements. In addition, the one-point reduced integration method is used in this method, which avoids subdomain integration for the discontinuous field with acceptable accuracy. However, this method is only applicable for the situation where the elements are only enriched with the Heaviside function, so cannot describe the crack tip accurately.

Menouillard et al. (2006) systematically investigated the stability of explicit integration in X-FEM and proposed a lumping scheme for the mass matrix associated with the additional degree of freedom. Fries and Belytschko (2006) combined X-FEM with a meshless method to avoid increasing additional unknowns. Ribeaucourt et al. (2007) considered the contact between the cracked surfaces. Jin et al. (2007) developed a cohesive crack model based on X-FEM and applied this model to simulate the cracking process of gravity dams. Zhuang and Cheng (2011b) applied X-FEM to study the propagation of subinterface cracks in bimaterials. Liu et al. (2011) integrated spectral element and X-FEM to improve the numerical dispersion in the dynamic fracture simulatiion.

Besides fracture mechanics, X-FEM has also been integrated into the other fields of mechanics and has produced fruitful works. Chessa and Belytschko (2004, 2006) extend the idea of X-FEM from the spatial scale to the time scale. Spatial-time X-FEM has unique advantages when handling problem with time discontinuities. Based on the same idea, Réthoré et al. (2005) developed a similar formulation and made remarkable progress. Sukumar et al. (2001) successfully applied X-FEM to simulate heterogeneous materials with voids and inclusions by constructing a new enrichment shape function, which was a significant advance in the area of modeling of composites. Moreover, Gracie et al. (2008) developed the X-FEM model, based on Volterra dislocations for a defect in crystalline material at the micro scale, and realized the FEM simulation of dislocation. In addition, its applications are growing in shear band evolution (Samaniego and Belytschko, 2005), multi-phase flows (Chessa and Belytschko, 2003a, b), mechanics of nano-interfaces (Farsad et al., 2010), multi-scale modeling (Belytschko et al., 2009) and the other subjects, which show broad application potential and lively life force.

1.3.2 Development of 3D X-FEM

The early study of X-FEM focused mainly on solving 2D fracture problems. With the rapid spread of its application, the 2D elements could not satisfy the needs of scientific research and engineering projects. Some complicated fracture problems, such as the collapse of buildings and bridges during earthquake, damage in the collision of vehicles, aircrafts and ships, instability propagation of cracks in pressure pipes, the fracture of mechanical components and so on, urgently require the development of 3D X-FEM.

X-FEM was first developed in 3D by Sukumar et al. (2000). For mode I crack problems, polar coordinates are established in the plane perpendicular to the crack tip, then the tip enrichment function is expressed in these coordinates and has the same formulations as 2D. The challenge in 3D fracture simulation is how to keep the crack surface and predicted crack path continuous and smooth. By adjusting the normal direction of the crack surface (Areias and Belytschko, 2005b), this condition can be approximately satisfied. Recently, Duan et al. (2009) constructed element local level set to describe the crack surface, and improved the continuity between crack surface and the predicted propagation direction by least square method. In addition, the crack branches, calculation of SIF and the dynamic propagation criterion are still the hot topics in the study of X-FEM on 3D simulations.

Areias and Belytschko (2005a) first introduced the concept of X-FEM into the shell element. They implemented the X-FEM formulation in four-node Belytschko—Lin—Tsay shell elements, and realized the arbitrary propagation of penetrated crack perpendicular to the middle surface of the shell. Because of the assumption in the Belytschko—Lin—Tsay shell element that the fiber along the thickness direction is not extendable, the change of thickness during deformation was not considered. For the simplicity of computer implementation, Areias and Belytschko (2006) also developed an adapted formulation-superposition of a pair of elements to deal with the crack in Kirchhoff—Love shell elements. In this new formulation, the displacement of the cracked element is the superposition of two newly created elements. It should be mentioned that there are three assumptions in the X-FEM shell elements above. (1) Normal constitutive law is used in the domain without cracks and cohesive law is used in the cracked region. These two domains are independent. (2) Cauchy stress is only related to the bounded terms of the deformation gradient. (3) The assumed deformation field is not related to the unbounded terms of the deformation gradient. A cohesive model is used to

FIGURE 1.6   X-FEM simulation of cracking in a pressure pipe (Song and Belytschko, 2009).

(a)   **(b)**

decide if the crack will propagate. The SIF is calculated by the superposition of plane membrane stress intensity factor K_m and plate bending stress intensity factor K_b. Using these methods, complex cracks can arbitrarily propagate in thin shells. Song and Belytschko (2009) successfully simulated crack propagation in pipelines, as shown in Figure 1.6.

Recently, based on continuum-based (CB) shell theory, Zhuang and Cheng (2011c) developed a novel X-FEM shell element, which has the following distinct advantages: the shell thickness variation and surface connection can be considered during deformation and fracture, so cracks not perpendicular to the mid-surface can be simulated; the stress intensity factors of the crack in the CB shell element can be calculated by using the "equivalent domain integral" method for 3D arbitrary nonplanar cracks. The continuum-based X-FEM shell is coded in the author's group with more than 10,000 Fortran statements and will be described in detail in Chapter 6.

1.4 ORGANIZATION OF THIS BOOK

Chapters 2 and 3 provide an introduction to fracture mechanics, to provide the essential concepts of static and dynamic linear elastic fracture mechanics, such as the crack propagation criterion, the calculation of stress intensity factor using the interaction integral, the nodal force release technique to simulate crack propagation in conventional FEM, and so on. These two chapters contain essential knowledge of fracture mechanics in preparation for the subsequent chapters. Readers who are familiar with fracture mechanics can skip these two chapters. Chapters 4 and 5 give the basic ideas and formulations of X-FEM. Chapter 4 focuses on the theoretical foundations, mathematical description of the enrichment shape function, discrete formulation, etc. In Chapter 5, based on the program developed by the authors and colleagues, numerical studies of two-dimensional fracture problems are provided to demonstrate the capability and efficiency of the algorithm and program of X-FEM in applications of strong and weak discontinuity problems. In Chapters 6–9, scientific research by the author's group is taken as examples to introduce the applications of X-FEM. In Chapter 6, a novel theory formula and computation method of X-FEM is developed for three-dimensional (3D) continuum-based (CB) shell elements to simulate arbitrary crack growth in shells using enriched shape functions. In Chapter 7, the algorithm is discussed and a program is developed based on X-FEM for simulating subinterfacial crack growth in bimaterials. Numerical analyses of crack growth in bimaterials gives a clear description of the effect on fracture made by the interface and loading. In Chapter 8, a method for representing discontinuous material properties in a heterogeneous domain by the X-FEM is applied to study ultrasonic wave propagation in polymer matrix particulate/fibrous composites. In Chapter 9, the simulation method of transient immiscible and incompressible two-phase flows is proposed, which demonstrates how to deal with the multi-phase flow problem by applying X-FEM methodology. Based on scientific research in the author's group, Chapter 10 gives the applications of X-FEM in the other frontiers of mechanics, e.g., nanomechanics, multi-scale simulations, and so on. In the Appendix, the essential concepts and methods of fracture mechanics, like the Westergaard stress function method and J integral, are provided in order to help the reader self-study.

Chapter 2

Fundamental Linear Elastic Fracture Mechanics

Chapter Outline
2.1 **Introduction**   **13**
2.2 **Two-Dimensional Linear Elastic Fracture Mechanics**   **15**
2.3 **Material Fracture Toughness**   **19**
2.4 **Fracture Criterion of Linear Elastic Material**   **20**
2.5 **Complex Fracture Criterion**   **22**
    2.5.1 Maximum Circumference Tension Stress Intensity Factor Theory   22
    2.5.2 Minimum Strain Energy Density Stress Intensity Factor Theory   24
    2.5.3 Maximum Energy Release Rate Theory   27
2.6 **Interaction Integral**   **29**
2.7 **Summary**   **31**

2.1 INTRODUCTION

Occurrences of fracture in solids arise predominantly from discontinuous surface displacement in the materials. The fracture problem is usually divided into three fundamental modes. The first of these is mode I opening cracking. The displacements are opposite and orientations are perpendicular to each other between two crack surfaces, which is a common form of crack in engineering practice, as shown in Figure 2.1(a). The second category is mode II sliding cracking, also called in-plane shear cracking. The displacements are also opposite between two crack surfaces; however, one displacement moves along and the other deviates from the crack growth direction, as shown in Figure 2.1(b). The third category is mode III antiplane shear cracking. The antiplane displacements occur between two crack surfaces, as shown in Figure 2.1(c).

During the fracture process, some energy is released at the crack tip location. The stress–strain field there must be related to the energy release

Extended Finite Element Method. http://dx.doi.org/10.1016/B978-0-12-407717-1.00002-9

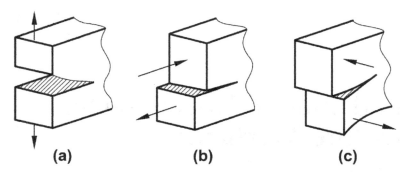

(a)   **(b)**   **(c)**

FIGURE 2.1   **Fracture modes.** (a) Opening crack. (b) Sliding crack. (c) Antiplane shear crack.

rate. If the intensity of the stress−strain field is strong enough, fracture occurs; otherwise cracking does not take place. Thus, the solution must be explored for the stress−strain field at the crack tip location. Modern fracture mechanics has provided some solutions using elastic mechanics analysis methods.

In real engineering problems, the loading conditions are complicated for general specimens. The location of crack initiation and direction of crack propagation are affected by the stress distribution. The crack tip location is usually under composite deformation conditions. For example, the inner surface crack in a pressure vessel forms an intersection angle with axial orientation. Under internal pressure, it is a mode I and II complex crack; once a penetrating crack or propagating crack forms, it will be a mode I, II, and III complex crack. Another example is rupture at an external edge of a wheel gear, which has a similar configuration to the partial eclipse of the moon; this is also a mode I and II complex crack. A further example is a fiber composite material plate, which is an anisotropic material. This type of crack shows zigzag growth, and is also a mode I and II complex crack. In order to set up the fracture criterion based on fracture mechanics, we pay more attention to whether crack propagation or arrest occurs after it is initiated, as well as crack velocity and direction. To determine crack propagation or arrest, it is necessary to establish the fracture criterion.

When using the idea of the extended finite element method (X-FEM) to construct a shape function at the crack tip location, it is necessary to have the aid of solutions from linear elastic fracture mechanics. Thus, 2D linear elastic fracture mechanics is briefly described in section 2.2, and the basic expressions of stress intensity factor are also provided. Material fracture toughness is discussed in section 2.3 and fracture criterion of linear elastic material is introduced in section 2.4. The computational method and fracture criterion for some complex fractures are provided in section 2.5. The interaction integral solution of stress intensity factor is given in section 2.6. In this chapter, the effect on the inertial force or kinetic energy for cracking is ignored, which is a form of quasi-steady-state fracture mechanics. Dynamic fracture mechanics will be introduced in Chapter 3.

2.2 TWO-DIMENSIONAL LINEAR ELASTIC FRACTURE MECHANICS

Griffith (1920) suggested that a crack growth depended on a criterion that related "the energy release of the structure to the surface area of a propagating crack". We consider an infinite plate with unit thickness subjected to uniaxial uniform tension, which is perpendicular to a central penetrating crack with length a. The strain energy of the plate is given by

$$V_\varepsilon = -\frac{\sigma^2}{2E}\pi a^2 \tag{2.1}$$

where E is elastic modulus. The surface energy to develop a crack is

$$E_S = 2\gamma a \tag{2.2}$$

where γ is surface energy on a unit area.

Crack propagation means that when the crack length increases by da, there is a balancing relation between the derivative of strain energy release to a crack length, which is a crack driving force, and the other derivative of crack surface energy with respect to a crack length, which is a resisting force preventing crack growth. That is,

$$\frac{\partial(E_S + V_\varepsilon)}{\partial a} = 0 \tag{2.3}$$

The external load for driving crack growth can be obtained by solving the above equations:

$$\sigma_f = \sqrt{\frac{2E\gamma}{\pi a}} \tag{2.4}$$

For ductile materials, the critical strain energy release rate G_C is usually used instead of surface energy rate, so it is given by

$$\sigma_f = \sqrt{\frac{EG_C}{\pi a}} \tag{2.5}$$

Example 2.1. An aluminum alloy cylinder pipeline, $G_C = 20$ N/mm, $E = 76$ GPa, has 300 MPa hoop stress induced by internal pressure. We want to find the crack growth length under this stress.

Solution. Using Eq. (2.5), the solution is obtained as

$$a = \frac{G_C E}{\pi \sigma^2} = \frac{20 \times 76 \times 10^3}{\pi \times 300^2} = 5.4 \text{ (mm)}$$

Obviously, the crack can be detected before it propagates to this distance. It is noted from this example that an important issue is to prevent the initial crack from happening and propagating. Here, we do not know if gas or liquid internal pressure results in the hoop stress on the pipe. If it is a gas with the compressible behavior, the potential damage is much more serious.

As well as the energy method, the stress intensity factor is another widely used method. For a 2D mode I crack problem as shown in Figure 2.2, we use a coordinate system with the crack tip as an original point. Here, the x-direction is the crack growth direction; the y-direction is the normal direction of the crack surface; the z-direction is the out-of-plane direction. If the stress intensity factor is known, analytical expressions for the displacement field and stress field can be solved at the crack tip location. Considering a stress element of a plane problem at the crack tip location with a polar coordinate system (r,θ), the analytical solution for a stress field of a plane mode I crack at the tip location is provided by the Westergaard stress function method, which is proved in Appendix A:

$$\sigma_x = \frac{K_{\mathrm{I}}}{\sqrt{2\pi r}}\cos\frac{\theta}{2}\left(1 - \sin\frac{\theta}{2}\sin\frac{3\theta}{2}\right)$$

$$\sigma_y = \frac{K_{\mathrm{I}}}{\sqrt{2\pi r}}\cos\frac{\theta}{2}\left(1 + \sin\frac{\theta}{2}\sin\frac{3\theta}{2}\right) \tag{2.6}$$

$$\tau_{xy} = \frac{K_{\mathrm{I}}}{\sqrt{2\pi r}}\cos\frac{\theta}{2}\sin\frac{\theta}{2}\cos\frac{3\theta}{2}$$

where K_{I} is called the **stress intensity factor** of the mode I crack. It is important to measure the stress intensity at the crack tip location. The

FIGURE 2.2 Stress element of the plane problem.

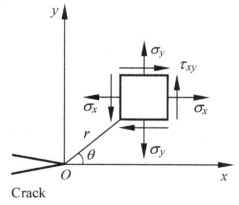

subscript I indicates that the crack is a mode I opening crack. Similarly, K_{II} and K_{III}, corresponding to mode II and III cracks, can be deduced. The plane stress field for a mode II crack at the tip location is given by

$$\sigma_x = -\frac{K_{II}}{\sqrt{2\pi r}}\sin\frac{\theta}{2}\left(2 + \cos\frac{\theta}{2}\cos\frac{3\theta}{2}\right)$$

$$\sigma_y = \frac{K_{II}}{\sqrt{2\pi r}}\sin\frac{\theta}{2}\cos\frac{\theta}{2}\cos\frac{3\theta}{2} \qquad (2.7)$$

$$\tau_{xy} = \frac{K_{II}}{\sqrt{2\pi r}}\cos\frac{\theta}{2}\left(1 - \sin\frac{\theta}{2}\sin\frac{3\theta}{2}\right)$$

In subsection 2.5.2, the stress field for a mode III crack at the tip location is presented as shown in Eq. (2.24).

When the stress intensity factor is known, the stress state at any point of the crack tip location with coordinates (r,θ) can be determined. For a mode I crack, the stress state near the crack tip is symmetrical to the crack and its elongation line. There is only normal stress at the elongation line, while the shear stress is zero; for a mode II crack, the normal stresses (σ_x, σ_y) near the crack tip are odd functions with respect to θ, while the shear stress (τ_{xy}) is an even function with respect to θ. Thus, the stress state near the crack tip is asymmetrical to the crack and its elongation line. There is only shear stress at the elongation line, while the normal stress is zero.

When a form of stress field at the crack tip location is constant, the strength can be fully determined by the value of the stress intensity factor. Because the material is elastic, the stress and displacement fields at the crack tip location can be found using elastic mechanics formulae. Here, it is noted that when $r \to 0$, at the crack tip point, the stress components may tend towards infinity. This is called **stress singularity**. It is geometrically discontinuous at the crack tip.

Since the stress, strain, displacement, and strain energy density at the crack tip location can be determined from the stress intensity factor for three fundamental crack modes, the stress intensity factor can be considered as an important parameter for characterizing the strength of the stress–strain field at the tip location. Modern fracture mechanics has benefited from the concept of the stress intensity factor proposed by Irwin (1957), which advanced the early stage fracture mechanics created by Griffith and set up the framework of linear elastic fracture mechanics.

In engineering applications, it is necessary to find the stress intensity factor. The computation method consists mainly of analytical and numerical methods. The former includes the stress function method and integration transformation method, and the latter includes the finite element method and so on. All of these methods require advanced mathematics and mechanics solutions, as well as complicated numerical computations. For convenience in engineering applications, for various forms of crack and loading and geometric dimensions, a large number of expression formulae of stress intensity factors are provided in references or manuals, whose dimension is [force][length]$^{-3/2}$. Some stress intensity factors for mode I cracks that are commonly used are listed in Table 2.1.

In Table 2.1, some commonly used stress intensity factors of mode I cracks are given for 2D infinite and semi-infinite plates. For 2D finite plates having a crack on the side of a hole, the stress intensity factor can be solved numerically. In general, the stress intensity factor is a physical quantity used to determine the entire stress field intensity at the crack tip location. The main purpose of linear elastic fracture mechanics is to determine the stress intensity factor for specimens with cracks.

TABLE 2.1 Some Stress Intensity Factors of Mode I Cracks

Case no.	Crack configuration	Stress intensity factor K_I
1	An infinite plate has a center crack with length $2a$ and is subjected to one-way uniform tension at two infinite ends (see Appendix A, Figure A.4)	$\sigma\sqrt{\pi a}$
2	An infinite plate has a center crack with length $2a$ and is subjected to uniform compression stress acting on the crack surfaces	$\sigma\sqrt{\pi a}$
3	An infinite plate has a center crack with length $2a$ and is subjected to one pair of compression forces F acting on the right side of crack surfaces and a remote distance b from the crack center (see Appendix A, Figure A.5)	$K_{\text{left}} = \frac{F}{\sqrt{\pi a}}\sqrt{\frac{a+b}{a-b}}$ $K_{\text{right}} = \frac{F}{\sqrt{\pi a}}\sqrt{\frac{a-b}{a+b}}$
4	A semi-infinite plate has one side crack with length a	$1.12\sigma\sqrt{\pi a}$
5	An infinite body has a circular crack with radius a	$2\sigma\sqrt{\frac{a}{\pi}}$
6	A plate strip with width b has a center crack with length $2a$	$\sigma\sqrt{b\tan\left(\frac{\pi a}{b}\right)}$
7	A plate strip with width b has two symmetric cracks, each of crack length a	$\sigma\sqrt{b\tan\left(\frac{\pi a}{b}\right) + 0.1\sin\left(\frac{2\pi a}{b}\right)}$

2.3 MATERIAL FRACTURE TOUGHNESS

The concept of stress intensity factor is introduced in section 2.2. The critical value of stress intensity factor K_{IC} is called **material fracture toughness**. A subscript I indicates a mode I crack. With reference to Table 2.1, the relation of crack length a, material fracture toughness K_{IC} and tension stress σ_f, at a location ahead of the crack tip and with enough energy to open the crack surfaces, can be described as follows:

$$\sigma_f = \frac{K_{IC}}{\alpha\sqrt{\pi a}} \tag{2.8}$$

where α is a geometrical parameter. For example, it is 1 for case no. 1 in Table 2.1. To obtain α values for different conditions, we can find them in a handbook, or can get them via analytical solution and numerical computation.

The concept of energy release and Griffith fracture criterion was introduced in section 2.1. If the strain energy release equation (2.5) satisfies the energy dissipation for crack growth, it is given by

$$\sigma_f = \sqrt{\frac{EG_{IC}}{\pi a}} \tag{2.9}$$

where G_{IC} is the energy release rate or crack driving force, which is compared with total strain energy before and after crack propagation. It should be noted that only mode I cracks are discussed here.

In fracture mechanics research, we pay close attention to the relation between energy release rate and stress intensity factor, because the former is the groundwork of the latter. They describe crack propagation or arrest from the point of view corresponding to the energy and stress fields. If we let $\alpha = 1$ in Eq. (2.8), substitute it into Eq. (2.9) and consider $E_1 = E/(1-v^2)$ for a plane strain problem, where v is Poisson's ratio, a relation can be found between material fracture toughness and energy release rate G_{IC} under plane strain conditions:

$$\frac{E}{1-v^2}G_{IC} = K_{IC}^2 \tag{2.10}$$

During the fracture process, the energy release is mainly consumed for material plastic flow at the crack tip location. For a specific material, the strain energy release rate required during the energy consumption process is called the **critical strain energy release rate**, G_{cr}. Thus, Eq. (2.9) may be rewritten as

$$\sigma_f = \sqrt{\frac{EG_{cr}}{(1-v^2)\pi a}} \tag{2.11}$$

This equation contains three aspect factors — material, stress level, and crack size — during the crack growth process.

The values of G_{IC} and K_{IC}, as well as their elastic modulus E, are listed in Table 2.2 for various typical materials. Since the material properties are

TABLE 2.2 Material Fracture Toughness

Material	G_{IC} (kN/m)	K_{IC} (MPa $\bullet$ m$^{1/2}$)	E (GPa)
Alloy steel	107	150	210
Aluminum alloy	20	37	69
Polymer	20	–	0.15
Low-carbon steel	12	50	210
Rubber	13	–	0.001
Thermal solidified reinforced glass	2	2.2	2.4
Polystyrene	0.4	1.1	3
Timber	0.12	0.5	2.1
Glass	0.007	0.7	70

different, these values vary greatly. For example, some polymers have higher fracture toughness, while stiffness alloy steel can prevent cracks from arbitrary propagation.

Example 2.2. An aluminum alloy pipe vessel is subjected to internal pressure with diameter 0.25 m, wall thickness 5 mm, material yield strength $\sigma_Y = 330$ MPa, material fracture toughness $K_{IC} = 41$ MPa$\sqrt{m}$, safety factor 0.75, and $\alpha = 1$. Determine the internal pressure for safety use and the maximum allowable crack length.

Solution. For the hoop stress of the vessel to reach the yield strength, the maximum internal pressure is given by

$$p = \frac{0.75\sigma\delta}{r} = \frac{0.75 \times 330 \times 0.005}{0.25/2} \text{ MPa} = 9.9 \text{ MPa}$$

From Eq. (2.8), the maximum allowable crack length is achieved:

$$a = \frac{K_{IC}^2}{\pi\sigma^2} = \frac{41^2}{\pi \times (0.75 \times 330)^2} \text{ m} = 8.74 \text{ mm}$$

We conclude that when a crack length is less than or equal to this value, the cracked vessel can be detected and repaired; when a crack is larger than this value, the crack may propagate unstably.

2.4 FRACTURE CRITERION OF LINEAR ELASTIC MATERIAL

At the crack tip location, the stress field has $r^{-1/2}$ singularity, as shown in Eqs. (2.6) and (2.7). If it is loading, the stress may tend towards infinity there.

From the viewpoint of traditional strength, no material could support this large a stress. Consequently, the structure must fail. However, this is not actually the case. A crack may propagate only when the loading reaches a particular value, and this results in the structure losing its load-bearing capability. We should note that the strength criterion established by using the stress as a control parameter cannot reflect the actual bearing capability for members containing the crack. Thus, it is necessary to choose a new control parameter and set up a new strength criterion. The failure criterion of fracture mechanics is different from the classical strength criterion, since it does not use the stress strength at a critical point and instead uses the stress intensity factor at the crack tip location as the fracture parameter (K_I, K_{II}, K_{III}). When the stress intensity factor is equal to or greater than the critical value (K_{cr}), the crack is in an initially unstable growth phase, which finally results in rupture.

What conditions of the members can experience unstable fracture? If it is a brittle material and a plastic area is relatively small enough at the crack tip, we only need one mechanics quantity to describe the stress and displacement fields at the crack tip location, which is the stress intensity factor K. Since K is related to energy release rate G, which is another mechanics quantity representing the capability for driving crack growth, the criterion of unstable fracture may be expressed as

$$K = K_{cr} \quad \text{or} \quad G = G_{cr} \tag{2.12}$$

where K_{cr} and G_{cr} represent the critical values of stress intensity factor and energy release rate respectively. This is the material fracture toughness. It is noted that $K \geq K_{cr}$ or $G \geq G_{cr}$ is used in some references rather than Eq. (2.12). The "greater than" symbol used only satisfies the computations in mathematics, but it has no physical meaning. Since the equals sign in Eq. (2.12) is enough to satisfy the condition of crack growth, which means there is enough energy to drive crack growth, Eq. (2.12) is therefore a necessary condition to denote crack propagation.

In the previous discussions, the high order terms of the stress field at the crack tip location are ignored, as presented in Eqs. (2.6) and (2.7) respectively. This is only suitable for a small area around the crack tip location, where it is called the K zone. The strength of the stress—strain field in the K zone can be measured by the stress intensity factor, while at the outside of the K zone, the high-order terms of the stress field have to be considered. Thus, it is not sufficient to use only the stress intensity factor as a unique quantity to measure the strength of the stress—strain field. If the size of the K zone is a few times larger than that of the fracture process zone, whether a crack occurs at this process zone is dependent on the intensity of the K zone outside of it; otherwise, if the size of the K zone is smaller than the fracture process zone, it is not meaningful to compute the stress intensity factor there. This is because it is not suitable for using macroscopic mechanics at the crack tip location. Thus, the fracture criterion may be set up under the critical strength condition at the K zone. Since the stress value does not reach infinity, there is always a plastic zone at the crack tip location and the stress in this zone is limited. We conclude that the condition for the established fracture criterion of the stress intensity factor is a plastic zone much smaller than the K zone, while the K zone is much smaller than the crack length. When a brittle crack occurs in many high-strength alloys and engineering materials, the intensity of the K zone always plays a dominant role. The fracture criterion of the stress intensity factor is suitable for brittle fracture of these materials.

2.5 COMPLEX FRACTURE CRITERION

2.5.1 Maximum Circumference Tension Stress Intensity Factor Theory

An infinite plate subjected to uniform loading includes a centric mode I and II complex crack. According to linear elastic fracture mechanics, the stress components can be written in the following forms at the crack tip location, based on Eqs. (2.6) and (2.7):

$$
\left.
\begin{aligned}
\sigma_x &= \frac{K_I}{\sqrt{2\pi r}}\cos\frac{\theta}{2}\left(1 - \sin\frac{\theta}{2}\sin\frac{3\theta}{2}\right) - \frac{K_{II}}{\sqrt{2\pi r}}\sin\frac{\theta}{2}\left(2 + \cos\frac{\theta}{2}\cos\frac{3\theta}{2}\right) \\
\sigma_y &= \frac{K_I}{\sqrt{2\pi r}}\cos\frac{\theta}{2}\left(1 + \sin\frac{\theta}{2}\sin\frac{3\theta}{2}\right) + \frac{K_{II}}{\sqrt{2\pi r}}\sin\frac{\theta}{2}\cos\frac{\theta}{2}\cos\frac{3\theta}{2} \\
\sigma_z &= \begin{cases} 0 & \text{plane stress} \\ 2\nu\dfrac{K_I}{\sqrt{2\pi r}}\cos\dfrac{\theta}{2} - 2\nu\dfrac{K_{II}}{\sqrt{2\pi r}}\sin\dfrac{\theta}{2} & \text{plane strain} \end{cases} \\
\tau_{xy} &= \frac{K_I}{\sqrt{2\pi r}}\cos\frac{\theta}{2}\sin\frac{\theta}{2}\cos\frac{3\theta}{2} + \frac{K_{II}}{\sqrt{2\pi r}}\cos\frac{\theta}{2}\left(1 - \sin\frac{\theta}{2}\sin\frac{3\theta}{2}\right)
\end{aligned}
\right\}
$$

$$(2.13)$$

where K_I and K_{II} are mode I and mode II stress intensity factors respectively.

Adopting the coordinate transformation, the above stresses in a right-angle coordinate system are transformed into circumference stresses in a polar coordinate system as given by

$$
\sigma_\theta = \frac{1}{\sqrt{2\pi r}}\cos\frac{\theta}{2}\left[\frac{K_I}{2}(1 + \cos\theta) - \frac{3K_{II}}{2}\sin\theta\right] \tag{2.14}
$$

where (r,θ) is a local polar coordinate system using the crack tip as an original point, as shown in Figure 2.2.

In maximum circumference tension stress intensity factor theory, the basic hypotheses are:

1. The crack opens along the orientation of the maximum circumference tension stress intensity factor, and is perpendicular to the direction of the maximum circumference tension stress.
2. When the maximum circumference tension stress intensity factor reaches a critical value, the crack propagates in an unstable manner. Obviously, this critical value is the material constant, which is the material fracture toughness, and has no relation to the ratio of K_I and K_{II}.

A circumference tension stress intensity factor K_θ is introduced as

$$K_\theta = \lim_{r \to 0} \sqrt{2\pi r}\sigma_\theta = \cos\frac{\theta}{2}\left[\frac{K_I}{2}(1 + \cos\theta) - \frac{3K_{II}}{2}\sin\theta\right] \tag{2.15}$$

Thus, Eq. (2.14) may be rewritten as

$$\sigma_\theta = \frac{1}{\sqrt{2\pi r}}K_\theta \tag{2.16}$$

From hypothesis 1, the opening angle θ can be determined through the attainment of the maximum value condition for Eq. (2.15), such that

$$\frac{\partial K_\theta}{\partial \theta}\Big|_{\theta=\theta_0} = 0, \qquad \frac{\partial^2 K_\theta}{\partial \theta^2}\Big|_{\theta=\theta_0} < 0, \tag{2.17}$$

Then we obtain:

$$\left.\begin{array}{l} K_I \sin\theta_0 - K_{II}(3\cos\theta_0 - 1) = 0 \\[2mm] K_I \cos\dfrac{\theta_0}{2}(1 - 3\cos\theta_0) + K_{II}\sin\dfrac{\theta_0}{2}(9\cos\theta_0 + 5) < 0 \end{array}\right\} \tag{2.18}$$

Combining the two equations from Eq. (2.18), the opening angle θ_0 can be determined.

From hypothesis 2, the fracture criterion is

$$K_{\theta\max} = \cos\frac{\theta_0}{2}\left[K_I \cos^2\frac{\theta_0}{2} - \frac{3K_{II}}{2}\sin\theta_0\right] = K_{\theta C} \tag{2.19}$$

The critical value $K_{\theta C}$ is a material fracture toughness. This is the fracture criterion for the maximum circumference tension stress intensity factor. An advantage is its simple form, but a disadvantage is that it cannot differentiate between the different situations of plane stress and plane strain.

For a pure mode I problem, let $K_{II} = 0$ in the first equation in Eq. (2.18), and $\theta_0 = 0$ is obtained. Then the fracture criterion of a pure mode I crack problem can be found via Eq. (2.19):

$$K_I = K_{\theta C} = K_{IC} \tag{2.20}$$

For a pure mode II problem, let $K_I = 0$ in the first equation in Eq. (2.18), then the opening angle is determined as

$$\theta_0 = -70.5° \tag{2.21}$$

as shown in Figure 2.3. Then, the fracture criterion of a pure mode II crack problem can be found via Eq. (2.19):

$$K_{II} = 0.87 K_{IC} = K_{IIC} \tag{2.22}$$

For an infinite plate subjected to uniform loading with a centric mode I and II complex crack, the displacement fields at the crack tip location are given by

$$u_x = \frac{K_I}{2\mu}\sqrt{\frac{r}{2\pi}}\cos\frac{\theta}{2}\left(\kappa - 1 + 2\sin^2\frac{\theta}{2}\right) + \frac{K_{II}}{2\mu}\sqrt{\frac{r}{2\pi}}\sin\frac{\theta}{2}\left(\kappa + 1 + 2\cos^2\frac{\theta}{2}\right)$$

$$u_y = \frac{K_I}{2\mu}\sqrt{\frac{r}{2\pi}}\sin\frac{\theta}{2}\left(\kappa + 1 - 2\cos^2\frac{\theta}{2}\right) - \frac{K_{II}}{2\mu}\sqrt{\frac{r}{2\pi}}\cos\frac{\theta}{2}\left(\kappa - 1 - 2\sin^2\frac{\theta}{2}\right)$$

$$\tag{2.23}$$

where μ is shear modulus.

2.5.2 Minimum Strain Energy Density Stress Intensity Factor Theory

According to linear elastic fracture mechanics, the stress components at the crack tip location can be found from Eq. (2.13) for the plane problem. For the

FIGURE 2.3 **Pure mode II crack problem.**

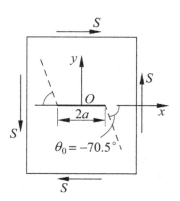

antiplane shear mode III crack problem, the stress intensity factor K_{III} can be obtained. Thus, the stress field at the crack tip location is given by

$$\tau_{zy} = \frac{K_{III}}{\sqrt{2\pi r}} \cos\frac{\theta}{2}$$

$$\tau_{zx} = -\frac{K_{III}}{\sqrt{2\pi r}} \sin\frac{\theta}{2}$$

(2.24)

where $K_{III} = S_y\sqrt{\pi a}$, which is dependent on the asymmetric antiplane load S_y and the crack length a.

The strain energy density W in an elastic body may be written as

$$W = \frac{1}{2E}\left(\sigma_x^2 + \sigma_y^2 + \sigma_z^2\right) - \frac{\nu}{E}\left(\sigma_x\sigma_y + \sigma_y\sigma_z + \sigma_z\sigma_x\right) + \frac{1}{2\mu}\left(\tau_{xy}^2 + \tau_{yz}^2 + \tau_{zx}^2\right)$$

(2.25)

Substituting Eqs. (2.13) and (2.24) into Eq. (2.25), an expression for strain energy density at the crack tip location can be obtained:

$$W = \frac{S}{r}$$

(2.26)

where S is called the **strain energy density factor:**

$$S = a_{11}K_I^2 + 2a_{12}K_IK_{II} + a_{22}K_{II}^2 + a_{33}K_{III}^2$$

(2.27)

and

$$a_{11} = \frac{1}{16\pi\mu}(1 + \cos\theta)(\kappa - \cos\theta)$$

$$a_{12} = \frac{1}{16\pi\mu}\sin\theta(2\cos\theta - \kappa + 1)$$

$$a_{22} = \frac{1}{16\pi\mu}[(\kappa + 1)(1 - \cos\theta) + (1 + \cos\theta)(3\cos\theta - 1)]$$

$$a_{33} = \frac{1}{4\pi\mu}$$

(2.28)

where

$$\kappa = \begin{cases} \dfrac{3 - \nu}{1 + \nu}, & \text{plane stress} \\[2mm] 3 - 4\nu, & \text{plane strain} \end{cases}$$

In minimum strain energy density stress intensity factor theory, the basic hypotheses are:

1. The crack opens along the orientation of the minimum strain energy density stress intensity factor.
2. When the minimum strain energy density stress intensity factor reaches a critical value, the crack propagates in an unstable manner. Obviously, this critical value is a material constant, which is the material fracture toughness, and has no relation to the ratio of K_I, K_{II}, and K_{III}.

From hypothesis 1, the opening angle θ should satisfy the attainment of the minimum value condition for Eq. (2.27):

$$\frac{\partial S}{\partial \theta}\Big|_{\theta=\theta_0} = 0, \quad \frac{\partial^2 S}{\partial \theta^2}\Big|_{\theta=\theta_0} > 0 \tag{2.29}$$

From hypothesis 2, the fracture criterion is

$$S_{\min} = S(\theta_0) = S_{\min C} \tag{2.30}$$

where $S_{\min C}$ is a critical value of the minimum strain energy density stress intensity factor, which is the material fracture toughness.

Adopting a similar method as described in section 2.5.1, for the plane strain problem we have

$$S_{\min C} = \frac{1-2\nu}{4\pi\mu} K_{IC}^2 \tag{2.31}$$

As a result, Eq. (2.30) has the following form:

$$K_{IC}^2 = \frac{4\pi\mu}{1-2\nu} \left[a_{11}K_I^2 + 2a_{12}K_I K_{II} + a_{22}K_{II}^2 + a_{33}K_{III}^2 \right] \tag{2.32}$$

In a pure mode II crack problem, $K_I = K_{III} = 0$, and we obtain from Eqs. (2.27)–(2.29):

$$\theta_0 = \arccos\left(\frac{1-2\nu}{3}\right) \tag{2.33}$$

Substituting Eq. (2.33) into Eq. (2.27), and considering Eq. (2.30), we get

$$K_{IIC} = \sqrt{\frac{3(1-2\nu)}{2(1-\nu)-\nu^2}} K_{IC} \tag{2.34}$$

If Poisson's ratio $\nu = 1/3$, then we have

$$\theta_0 = -83°37', \quad K_{IIC} = 0.9K_{IC} \tag{2.35}$$

This result is very close to $K_{IIC} = 0.87K_{IC}$, which is given by Eq. (2.22) from the fracture criterion of the maximum circumference tension stress intensity factor, as given in subsection 2.5.1.

2.5.3 Maximum Energy Release Rate Theory

From the viewpoint of physics, the orientation of crack growth must be in the direction of the maximum energy released. For a plane problem, the polar coordinate forms of stress components at the crack tip location are

$$\sigma_\theta = \frac{1}{2\sqrt{2\pi r}} \cos\frac{\theta}{2} [K_I(1 + \cos\theta) - 3K_{II}\sin\theta]$$

$$\sigma_r = \frac{1}{2\sqrt{2\pi r}} \cos\frac{\theta}{2} [K_I \sin\theta + K_{II}(3\cos\theta - 1)]$$

$$(2.36)$$

and we introduce

$$K_{1\theta} = \frac{1}{2}\cos\frac{\theta}{2} [K_I(1 + \cos\theta) - 3K_{II}\sin\theta]$$

$$K_{2\theta} = \frac{1}{2}\cos\frac{\theta}{2} [K_I \sin\theta + K_{II}(3\cos\theta - 1)]$$

Then, Eq. (2.36) is rewritten as

$$\sigma_\theta = \frac{K_{1\theta}}{\sqrt{2\pi r}}, \qquad \sigma_r = \frac{K_{2\theta}}{\sqrt{2\pi r}} \tag{2.37}$$

The relation between stress intensity factor K_{IC} and energy release rate G_{IC} given in Eq. (2.10) for a plane strain problem can be extended to polar coordinates. The energy release rate G_θ for crack propagation along the θ direction is deduced by

$$G_\theta = \frac{1 - \nu^2}{E} \left(K_{1\theta}^2 + K_{2\theta}^2\right) \tag{2.38}$$

For the energy release rate criterion, the basic hypotheses are:

1. The crack propagates along the orientation of the maximum energy release rate, such that

$$\frac{\partial G_\theta}{\partial \theta}\Big|_{\theta=\theta_0} = 0, \quad \frac{\partial^2 G_\theta}{\partial \theta^2}\Big|_{\theta=\theta_0} < 0 \tag{2.39}$$

2. When the maximum energy release rate reaches a critical value, the crack propagates unstably. That is,

$$G_{\theta\max} = G_\theta(\theta_0) = G_{\theta C} \tag{2.40}$$

This critical value is a material constant, which is the material fracture toughness, and has no relation to the ratio between $K_{1\theta}$ and $K_{2\theta}$.

According to hypothesis 1, the opening angle θ_0 may be obtained by the following equation:

$$K_I^2 \sin \theta_0 (1 + \cos \theta_0) - 2K_I K_{II} (\sin^2 \theta_0 - \cos^2 \theta_0 - \cos \theta_0)$$
$$+ K_{II}^2 \sin \theta_0 (1 - 3 \cos \theta_0) = 0 \tag{2.41}$$

From hypothesis 2, the fracture criterion should be given by

$$G_{\theta max} = \frac{1 - \nu^2}{4E} (1 + \cos \theta_0) \left[K_I^2 (1 + \cos \theta_0) - 4K_I K_{II} \sin \theta_0 \right.$$
$$\left. + K_{II}^2 (5 - 3 \cos \theta_0) \right] = G_{\theta C} \tag{2.42}$$

Using Eq. (2.10), which is a relation between stress intensity factor and energy release rate for a mode I plane strain crack, we have

$$G_{\theta C} = \frac{1 - \nu^2}{E} K_{IC}^2 \tag{2.43}$$

Accordingly, Eq. (2.42) becomes

$$K_{IC}^2 = \frac{1}{4}(1 + \cos \theta_0) \left[K_I^2 (1 + \cos \theta_0) - 4K_I K_{II} \sin \theta_0 + K_{II}^2 (5 - 3 \cos \theta_0) \right] \tag{2.44}$$

This is the maximum energy release rate criterion.

In engineering practice, the simpler and more conservative criteria are used for unstable crack growth. Equation (2.38) can be modified as

$$K_I^2 + K_{II}^2 \geq K_{IC}^2 \tag{2.45}$$

A more conservative alternative is

$$K_I + K_{II} \geq K_{IC} \tag{2.46}$$

The above two criteria are illustrated in Figure 2.4. The former has a safety margin within a quarter circle and the latter has a safety margin within a triangle connected inside a quarter circle. Obviously, the latter is more conservative than the former.

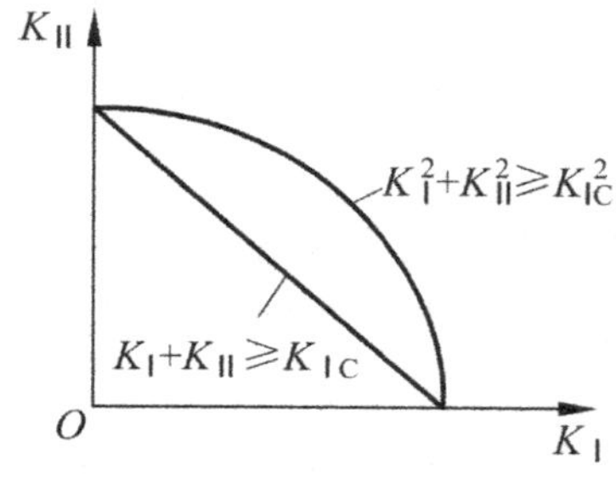

FIGURE 2.4 Engineering complex fracture criteria.

A similar process may deduce the engineering fracture criteria for three-dimensional complex cracks and the safety zone is an interior within a spherical or elliptic ball.

2.6 INTERACTION INTEGRAL

As introduced in previous sections, the stress intensity factor is an important parameter to estimate crack growth. In the computation of fracture mechanics, J integration is adopted to calculate the stress—deformation field at the crack tip location. If two group deformation fields exist in a domain, J integration is given by

$$J^{(1+2)} = \int_{\Gamma} \left[\frac{1}{2} \left(\sigma_{ij}^{(1)} + \sigma_{ij}^{(2)} \right) \left(\varepsilon_{ij}^{(1)} + \varepsilon_{ij}^{(2)} \right) \delta_{1j} - \left(\sigma_{ij}^{(1)} + \sigma_{ij}^{(2)} \right) \frac{\partial \left(u_i^{(1)} + u_i^{(2)} \right)}{\partial x_1} \right] n_j \mathrm{d}\Gamma$$

(2.47)

This equation can be written in a more concise form:

$$J^{(1+2)} = J^{(1)} + J^{(2)} + I^{(1+2)}$$

(2.48)

where $J^{(1)}$ or $J^{(2)}$ correspond to deformation field 1 or field 2. $I^{(1+2)}$ is called the interaction integral. If it is linear elastic, then $J = K$, which is a stress intensity factor.

In the extended finite element method, the interaction integral is often used to find the stress intensity factor. The interaction integral is an energy integral in a loop Γ including the crack tip, as shown in Figure 2.5. The definition of interaction integral is (Moës et al., 1999):

$$I^{(1+2)} = \int_{\Gamma} \left[W^{(1,2)} \delta_{1j} - \sigma_{ij}^{(1)} \frac{\partial u_i^{(2)}}{\partial x_1} - \sigma_{ij}^{(2)} \frac{\partial u_i^{(1)}}{\partial x_1} \right] n_j \mathrm{d}\Gamma$$

(2.49)

where $(\sigma_{ij}^{(1)}, \varepsilon_{ij}^{(1)}, u_i^{(1)})$ are variables in the actual stress—deformation field and $(\sigma_{ij}^{(2)}, \varepsilon_{ij}^{(2)}, u_i^{(2)})$ are variables in the auxiliary stress—deformation field. The

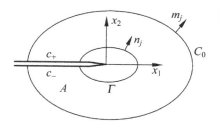

FIGURE 2.5   Schematic of interaction integral.

auxiliary field is an arbitrary stress–deformation field satisfying the deformation formula and constitutive equation. By the following deduction, we explain how to choose the suitable auxiliary field and find the solution.

When a loop Γ is close to the crack tip, there is a relation between the stress intensity factors in the actual and auxiliary deformation fields respectively, and interaction integral (Moës et al., 1999):

$$I^{(1+2)} = \frac{2}{E^*}\left(K_{\mathrm{I}}^{(1)}K_{\mathrm{I}}^{(2)} + K_{\mathrm{II}}^{(1)}K_{\mathrm{II}}^{(2)}\right) \tag{2.50}$$

where E^* is a combination term of material constant E (Young's modulus) and ν (Poisson's ratio):

$$E^* = \begin{cases} E & \text{plane stress} \\ \dfrac{E}{1-\nu^2} & \text{plane strain} \end{cases}$$

It can be seen from the equation above that, if the auxiliary field satisfies $K_{\mathrm{I}}^{(2)} = 1$, $K_{\mathrm{II}}^{(2)} = 0$, the stress intensity factor $K_{\mathrm{I}}^{(1)}$ for a mode I crack in the actual stress–deformation field can be solved directly using the interaction integral equation (2.50). Also, letting $K_{\mathrm{I}} = 1$ in Eq. (2.6),

$$\sigma_x = \frac{1}{\sqrt{2\pi r}}\cos\frac{\theta}{2}\left(1 - \sin\frac{\theta}{2}\sin\frac{3\theta}{2}\right)$$

$$\sigma_y = \frac{1}{\sqrt{2\pi r}}\cos\frac{\theta}{2}\left(1 + \sin\frac{\theta}{2}\sin\frac{3\theta}{2}\right) \tag{2.51}$$

$$\tau_{xy} = \frac{1}{\sqrt{2\pi r}}\cos\frac{\theta}{2}\sin\frac{\theta}{2}\cos\frac{3\theta}{2}$$

Using the same principle, if we want to find the stress intensity factor $K_{\mathrm{II}}^{(1)}$ for a mode II crack in the actual stress–deformation field, we may only choose the auxiliary field to satisfy $K_{\mathrm{I}}^{(2)} = 0$, $K_{\mathrm{II}}^{(2)} = 1$, which can also be found by letting $K_{\mathrm{II}} = 1$ in Eq. (2.7),

$$\sigma_x = -\frac{1}{\sqrt{2\pi r}}\sin\frac{\theta}{2}\left(2 + \cos\frac{\theta}{2}\cos\frac{3\theta}{2}\right)$$

$$\sigma_y = \frac{1}{\sqrt{2\pi r}}\sin\frac{\theta}{2}\cos\frac{\theta}{2}\cos\frac{3\theta}{2} \tag{2.52}$$

$$\tau_{xy} = \frac{1}{\sqrt{2\pi r}}\cos\frac{\theta}{2}\left(1 - \sin\frac{\theta}{2}\sin\frac{3\theta}{2}\right)$$

The displacement fields are obtained similarly.

In practical applications, the interaction integral equation (2.49) is usually transformed into an area integral for convenient calculation, as shown in

Figure 2.5. The other loop C_0 is set up beyond the loop Γ. The area between these two loops is A. Then, the interaction integral can be written as

$$I^{(1+2)} = \int_A \left[W^{(1,2)} \delta_{1j} - \sigma_{ij}^{(1)} \frac{\partial u_i^{(2)}}{\partial x_1} - \sigma_{ij}^{(2)} \frac{\partial u_i^{(1)}}{\partial x_1} \right] q m_j \mathrm{d}\Gamma \qquad (2.53)$$

where q is a smooth weight function, which becomes 0 at C_0 and 1 at Γ. m_j is outer normal on the loop C_0.

2.7 SUMMARY

In this chapter, the concept of energy release and balance during the crack growth process has been introduced. The stress fields are provided for mode I, II, and III cracks at the crack tip location. In the subsequent chapters, these solutions of stress—strain fields can be used as the basis of enriched shape functions to construct the enriched field of extended finite elements at the crack tip location.

Based on the analysis of linear elastic theory, the concepts of stress intensity factor and material fracture toughness have been introduced. The analytical method is given to determine the stress intensity factor. The failure criterion and design standard are set up for linear elastic fracture mechanics. The stress intensity factor K is a unique quantity to describe the intensity of a crack tip field, which is an important parameter in determining brittle material fracture. For a specific type of fracture, the crack tip field can be fully determined using a parameter K, which can be used to deduce the energy release rate.

Next, three kinds of complex fracture theories are discussed: the maximum circumference tension stress intensity factor theory, the minimum strain energy density stress intensity factor theory, and the maximum energy release rate theory. The crack orientation angle, fracture criterion, and stress intensity factor of complex mode cracks can be found using this analytical method. Finally, a method using the interaction integral is employed to find the stress intensity factor under quasi-static conditions.

Chapter 3

Dynamic Crack Propagation

Chapter Outline

3.1 Introduction to
Dynamic Fracture
Mechanics   **33**
3.2 Linear Elastic Dynamic
Fracture Theory   **35**
  3.2.1 Dynamic Stress
Field at Crack Tip
Position   35
  3.2.2 Dynamic Stress
Intensity Factor   37
  3.2.3 Dynamic Crack
Propagating
Condition and
Velocity   38

3.3 Crack Driving Force
Computation   **41**
  3.3.1 Solution Based on
Nodal Force Release   41
  3.3.2 Solution Based on
Energy Balance   43
3.4 Crack Propagation in
Steady State   **44**
3.5 Engineering Applications of
Dynamic Fracture
Mechanics   **45**
3.6 Summary   **49**

3.1 INTRODUCTION TO DYNAMIC FRACTURE MECHANICS

Dynamic fracture is used to describe the crack growth processes accompanied by rapidly occurring changes in loading and crack/structure geometry, where these cases are not well described by a sequence of static equilibrium states. The inertial forces must be included in the equations of motion of the body, so the term "dynamic fracture mechanics" can be applied to the problem. Landmark works in this area were contributed by a physicist (Mott, 1948), who studied two general classes of problems: stationary cracks or initiation under dynamic loading and rapid crack propagation or arrest following initiation. A key subsequent step was identified by Irwin (1969), who proposed six research areas — crack speed, crack direction, crack division, crack arrest, the onset of rapid fracture, and minimum toughness — and introduced the concept of progressive fracture to link these areas. This development ushered in the modern approach to dynamic fracture

Extended Finite Element Method. http://dx.doi.org/10.1016/B978-0-12-407717-1.00003-0

mechanics. In the 1980s, a more extensive definition was given by Kanninen and Popelar (1985): dynamic fracture mechanics includes all fracture mechanics problems for rapidly changing loads and crack geometry. All of the problems relating to the time domain and space domain of crack rapid initiation, propagation, and arrest are within the scope of dynamic fracture mechanics.

A principal difference between dynamic and static states is whether the quantity is affected by inertial force. The characterization quantity is a ratio of kinetic energy occupied within the internal energy. If there is the same quantity of kinetic energy and strain energy, the proportion of kinetic energy is not ignored in the internal energy formulation, which is a dynamic problem. Simulation analysis of crack propagation in polyethylene gas pipelines, with pipe diameter 250 mm, wall thickness 22.7 mm, and internal pressure 0.311 MPa, in terms of energy variation with time, is illustrated in Figure 3.1, with assumed crack propagation velocity of 200 m/s. There are intermediate values between kinetic energy and strain energy, which are combined within the internal energy. The error between external work and internal energy indicates that the energy is no longer conservative during the crack propagation process, which is in terms of the energy balance situation. The energy error is reduced to form the new crack surfaces. All rapid crack growth events are dynamic fracture problems.

The research goal on crack initiation is to find the reason for fracture initiation. Actually, it involves only a part of the fracture process, while research for crack propagation and arrest is looking for a second line of defense against catastrophic rupture if crack initiation cannot be prevented. It clearly consists of a process. One of the main differences compared with the crack initiation problem is that the crack size is an unknown function that varies with time. Since a propagation crack produces the displacement boundary condition of a structure, the moving boundary problem is a

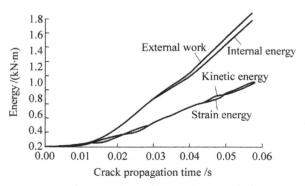

FIGURE 3.1 **Energy variation during crack propagation in a polyethylene pipeline.**

nonlinear problem although the control equations are linear. The balance equation is linked to a momentum equation since the time variable increases.

Because of material inertia, the load propagates in the form of a stress wave. Then, crack growth produces new free boundaries. The interactions between the stress wave and crack make the dynamic fracture problem more complicated than the static crack case. For example, the static tension load, which is parallel to the crack direction, does not produce a concentration force at the crack tip location, while a tension stress impulse wave in the same direction will cause a mode I crack to occur as it propagates to the tip location. This is due to a transverse inertial effect. On the other hand, the dynamic material fracture toughness is related to the loading rate. This increases the difficulty of analysis and experiments for dynamic fracture problems because of high stress gradients, loading stress wave forms, and rapid crack propagation at the crack tip location.

3.2 LINEAR ELASTIC DYNAMIC FRACTURE THEORY

3.2.1 Dynamic Stress Field at Crack Tip Position

Under dynamic loading, the expression of the stress field for a linear elastic material at the crack tip location is given by

$$\sigma \approx \frac{K(t)}{\sqrt{2\pi r}} f(\theta, \dot{a}) \tag{3.1}$$

If the crack speed $\dot{a}$ is zero in Eq. (3.1), we have the same situation as in a quasi-steady state. The angle distributed function $f(\theta)$ is also the same, as in Eqs. (2.6) and (2.7). Therefore, the stress intensity factor K can also be used to express this dynamic field. The critical value is the dynamic material fracture toughness. Another quantity used to describe the field at the crack tip is the energy release rate G, which is also expressed in terms of crack driving force.

For a propagation crack, the approaching solution of stress field at crack tip has $r^{-1/2}$ singularity. The angle distribution function is related to crack velocity $\dot{a}$, so the stress intensity factor can also be used to describe the stress situation at the crack tip location. For a mode I crack, the relation of G_I and K_I is

$$G_I = A(\dot{a}) K_I^2 / E \tag{3.2}$$

where $A(\dot{a})$ is a monotonously increasing progressive function.

For the more generic problem, we have the situation of a semi-infinite mode I crack under nonconstant loading and crack growth at nonconstant speed. The stress intensity factor at the crack tip is only related to loading history, current crack position l, and instantaneous velocity of crack propagation. For a mode I crack, it can be expressed as

$$K_I(t, l, \dot{a}) = K(\dot{a})K_I(t, l, 0) \tag{3.3}$$

where $K_I(t,l,0)$ is the stress intensity factor produced by the same load at the crack, which has the same current crack length. $K(\dot{a})$ is a factor related to the instantaneous crack propagation velocity; that is,

$$K(\dot{a}) \approx (1 - \dot{a}/C_r)/\sqrt{1 - \dot{a}/C_d} \tag{3.4}$$

where C_r and C_d are **Rayleigh wave speed** and **expansion wave speed** respectively. As can be seen from Eq. (3.4), the critical speed of crack growth is Rayleigh wave speed. If the crack speed is lower than this value, it is called sub-Rayleigh wave speed, at which the crack commonly propagates. A sub-sound speed crack is usually used to represent a sub-Rayleigh wave speed crack. A body wave, which does not cause distortion, propagates at the expansion wave speed C_d. This type of wave is also called a longitudinal wave or unspin wave. The body wave, which does not cause a volume change, propagates at speed C_s. This type of wave is also called a transverse wave, an equal-volume wave or a distortion wave, and is dependent on the material elastic constants, given by

$$C_d = \sqrt{\frac{\lambda + 2\mu}{\rho}}, \quad C_s = \sqrt{\frac{\mu}{\rho}} \tag{3.5}$$

where λ and μ are Lamé constants, and ρ is material density. It is known that the ratio of them is a function of Poisson's ratio.

Rayleigh wave speed C_r can be solved by finding the root of the Rayleigh function as given below:

$$R\left(\frac{1}{C_r}\right) = \left(2\frac{1}{C_r^2} - \frac{1}{C_s^2}\right)^2 + 4\frac{1}{C_r^2}\left(\frac{1}{C_d^2} - \frac{1}{C_r^2}\right)^{1/2}\left(\frac{1}{C_s^2} - \frac{1}{C_r^2}\right)^{1/2} = 0 \tag{3.6}$$

Rayleigh wave speed is slightly lower than transverse wave speed, while longitudinal wave speed is higher than transverse wave speed. As an example, when Poisson's ratio $\nu = 1/3$, $C_r/C_s \approx 0.9$ and $C_d/C_s = 2$ respectively.

For a finite body, when a stress wave reflected from a crack surface is reflected again to the crack from the boundary, the solution is no longer valid. The additional load produced by the finite boundary is not only related to external load and specimen geometry and configuration, but also to crack propagation history. In the following sections, we will describe how to solve the dynamic stress intensity factor under general conditions.

3.2.2 Dynamic Stress Intensity Factor

Dynamic stress intensity factors K_I and K_{II} can be defined via the progressive behavior of a stress field at the crack tip position (Freund, 1990):

$$K_I = \lim_{r \to 0} \sqrt{2\pi r \sigma_{yy}}, \quad K_{II} = \lim_{r \to 0} \sqrt{2\pi r \sigma_{xy}} \tag{3.7}$$

The relation between energy release rate and stress intensity factor is given by

$$G = \frac{1 - \nu^2}{E} \left[\beta_1(\dot{a}) K_I^2 + \beta_2(\dot{a}) K_{II}^2 \right] \tag{3.8}$$

where E is Young's modulus, ν is Poisson's ratio, and β_i ($i = 1, 2$) is a function of crack propagation velocity $\dot{a}$:

$$\beta_i(\dot{a}) = \frac{\dot{a}^2 \alpha_i}{(1 - \dot{a}) c_2^2 D} \tag{3.9}$$

$$\alpha_i = \sqrt{1 - \frac{\dot{a}^2}{c_i^2}}, \quad D(\dot{a}) = 4\alpha_1\alpha_2 - \left(1 + \alpha_2^2\right)^2 \tag{3.10}$$

and where c_1 and c_2 are longitudinal wave speed and transverse wave speed respectively, provided by Eq. (3.5).

As it is similar to solving the steady state stress intensity factor described in Chapter 2, K_I and K_{II} can be obtained by means of the interaction integral I_{int} through an auxiliary field and virtual crack growth direction q (Attigui and Petit, 1997):

$$\begin{aligned} I_{\text{int}} = &- \int_\Omega (\sigma^{\text{aux}} : \nabla u - \rho \ddot{u} u^{\text{aux}}) \text{div}(q) d\Omega + \int_\Omega \sigma^{\text{aux}} : (\nabla u \nabla q) + \sigma : (\nabla u^{\text{aux}} \nabla q) d\Omega \\ &+ \int_\Omega \text{div}(\sigma^{\text{aux}}) \nabla u(q) + \rho \ddot{u} \nabla u^{\text{aux}} d\Omega + \int_\Omega \rho \ddot{u}^{\text{aux}} \nabla \dot{u}(q) + \rho \dot{u} \nabla \dot{u}^{\text{aux}}(q) d\Omega \end{aligned} \tag{3.11}$$

where σ^{aux} and u^{aux} are the auxiliary stress field and displacement field respectively, and q is parallel to the crack surface, which is defined by

$$q = \begin{cases} 0 & \text{outside of surface } S_1 \cup S_2 \\ 1 & \text{inside of surface } S_1 \\ \text{linear variation} & \text{inside of surface } S_2 \end{cases} \tag{3.12}$$

where S_1 and S_2 denote areas that are defined close to the crack tip, as shown in Figure 3.2. The arrowheads are used to represent the quantities and orientations of vector q inside the surfaces S_1 and S_2.

The integral is related to the stress intensity factor as

$$I_{\text{int}} = \frac{2(1 - \nu^2)}{E} \left[\beta_1(\dot{a}) K_I K_I^{\text{aux}} + \beta_2(\dot{a}) K_{II} K_{II}^{\text{aux}} \right] \tag{3.13}$$

FIGURE 3.2   Schematic of interaction integral area.

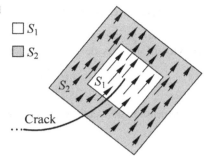

By selecting a suitable auxiliary displacement field u^{aux}, the various stress intensity factors can be obtained: for example, let $K_{\text{I}}^{\text{aux}} = 1$ and $K_{\text{II}}^{\text{aux}} = 0$, then K_{I} can be determined; in turn, using $K_{\text{I}}^{\text{aux}} = 0$ and $K_{\text{II}}^{\text{aux}} = 1$, K_{II} can be obtained. In the following sections on extended finite element analysis, we will use this method to calculate the dynamic stress intensity factor.

3.2.3 Dynamic Crack Propagating Condition and Velocity

Crack initiation and propagation occur in a fractured structure under dynamic loading, resulting in structure failure or crack arrest. The condition of these processes is called a criterion. The bulk of current state-of-the-art applications of dynamic fracture mechanics to rapid crack propagation and arrest are based upon the use of the concept of dynamic crack arrest. This point of view, which gives direct consideration to crack propagation with crack arrest occurring only when continued propagation becomes impossible, is in good agreement with observations. Within the confines of elastic dynamic behavior, rapid crack propagation or arrest occurs in such an approach under the condition that

$$G_{\text{max}} = G_{\text{d}}(T, \dot{a}, h) \qquad \text{Crack propagation} \qquad (3.14)$$

$$G < G_{\text{dmin}} \qquad \text{Crack arrest} \qquad (3.15)$$

where G, the crack driving force, is a function of the crack geometry, structure dimensions, and load intensity. It is calculated from some analytical models whilst G_{d}, the dynamic fracture resistance of the material, determined experimentally, is a function of the temperature T, crack speed $\dot{a}$, and thickness of the material h. Using this relation, if continued crack propagation requires that $G = G_{\text{d}}$ (Eq. (3.14)), then it is clear that crack arrest will occur at a time when G becomes less than the minimum value of G_{d} for all subsequent times, as shown in Eq. (3.15). Therefore, to prevent crack propagation rupture, the relative values of G and G_{d} provide a quantity condition for real applications. In Eqs. (3.14) and (3.15), the relation of G and G_d produces the energy criterion of crack growth. An alternative form of Eq. (3.2) uses K and K_d. Similarly, the relation of crack tip opening angle CTOA and $(CTOA)_C$ comprises another geometry criterion to evaluate crack growth.

There are two kinds of analytical modes developed: (a) generation mode; (b) propagation mode. The generation mode calculates the crack driving force G when the geometric dimensions and working conditions of the structure are provided, which is a widely used method via crack propagation equation (3.14) to evaluate the material fracture toughness G_d. The propagation mode measures crack velocity and pressure distribution on the structure using crack propagation experiments. Then, the velocity, pressure, as well as material and geometry conditions in the experiment are used to compute G_d, which is equivalent to the crack driving force G from Eq. (3.14). Now, finite element software has been developed to compute G and enable simulation of crack propagation (Zhuang, 2000a).

In the general analysis, crack velocity is measured experimentally. Mott (1948) deduced crack propagation velocity in an infinite plate by using a dimensionless analysis method, which assumes that the in-plane displacement components in the plate are given by

$$
\left.
\begin{aligned}
u &= c_1 \sigma\, a/E \\
v &= c_2 \sigma\, a/E
\end{aligned}
\right\}
\tag{3.16}
$$

where c_1 and c_2 are dimensionless ratios, which are not explicit functions of time. If it is necessary to solve the velocity components, the first derivative of displacement with respect to time is required. Equation (3.16) becomes

$$
\left.
\begin{aligned}
\dot{u} &= c_1 \sigma\, \dot{a}/E \\
\dot{v} &= c_2 \sigma\, \dot{a}/E
\end{aligned}
\right\}
\tag{3.17}
$$

From the kinetic energy definition, we have

$$
E_K = \frac{1}{2}\rho \int \left(\dot{u}^2 + \dot{v}^2 \right) dxdy
\tag{3.18}
$$

where ρ is material density. Substituting Eq. (3.17) into Eq. (3.18), the kinetic energy is obtained as

$$
E_K = \frac{1}{2}\rho\, \dot{a}^2 \frac{\sigma^2}{E^2} \int \left(c_1^2 + c_2^2 \right) dxdy
\tag{3.19}
$$

Since we are considering an infinite plate, the unique geometrical parameter is crack growth length a. The integral in Eq. (3.19) must be proportional to a^2. Introducing a dimensionless constant k, we then rewrite Eq. (3.19) as follows:

$$E_K = \frac{1}{2} k \rho a^2 \dot{a}^2 \frac{\sigma^2}{E^2} \tag{3.20}$$

Another expression for kinetic energy is

$$E_K = \int_{a_0}^{a} (G - R) \mathrm{d}a \tag{3.21}$$

where the lower limit of the integral is a crack semi-length a_0 at crack initiation, G is the dynamic crack driving force, and R is resistance. At a critical point of unstable fracture, from Eq. (2.5) we have

$$G = G_{IC} = \frac{\pi \sigma^2 a_0}{E} \tag{3.22}$$

It is supposed that σ is unchanged after crack propagation. If the resistant curve and dynamic effect are not considered, the resistance is $R = G_{IC}$ and $G = \pi \sigma^2 a / E$ given by Eq. (3.22). This equation is only suitable for situations where crack velocity $\dot{a}$ is far less than acoustic speed. Therefore, the total kinetic energy of a plate is obtained as

$$E_K = 2 \int_{a_0}^{a} (G - R) \mathrm{d}a = \frac{\pi \sigma^2}{E} (a - a_0)^2 \tag{3.23}$$

From Eqs. (3.20) and (3.23), we have

$$\dot{a} = \sqrt{\frac{2\pi}{k}} \sqrt{\frac{E}{\rho}} \left(1 - \frac{a_0}{a} \right) \tag{3.24}$$

where $\sqrt{E/\rho} = C_0$ is the one-dimensional propagation speed of an elastic wave, which is the acoustic speed. For brittle fracture, the value of $\sqrt{2\pi/k}$ is approximately equivalent to 0.38 based on experiments. Thus, by experiments and analysis, an expression for mode I crack propagation velocity, Eq. (3.24), can be written as

$$\dot{a} = 0.38 C_0 \left(1 - \frac{a_0}{a} \right) \tag{3.25}$$

The crack propagation rates of some materials are listed in Table 3.1. The higher the material toughness, the lower the crack velocity. The crack velocity may be more than 1000 m/s if brittle fracture occurs in steel in a low-temperature environment. It is thus clear that the damage is very serious.

TABLE 3.1 Crack Propagation Rates of Materials

Material	C_0 (m/s)	$\dot{a}$ (m/s)	$\dot{a}/C_0$
Glass	5200	1500	0.29
Steel (brittle fracture)	5000	1000–1400	0.20–0.28
Man-made fiber	1100	400	0.37

For pressurized steel pipelines, Kanninen and Popelar (1985) have proposed a computation formula for crack velocity:

$$\dot{a} = \frac{3}{4}C_0\left(\frac{h}{r}\right)^{1/2} \tag{3.26}$$

where C_0 is the speed of an elastic wave propagating in steel, which is approximately 5000 m/s; h and r are wall thickness and external radius of the pipe respectively. For example, for a pipe of thickness 8 mm and diameter 720 mm, the crack velocity $\dot{a}$ is found to be approximately 550 m/s calculated by Eq. (3.26). Using this velocity, we can compute the crack driving force and estimate the fracture toughness.

3.3 CRACK DRIVING FORCE COMPUTATION

There are currently only a small number of analytical solutions adopted to analyze dynamic crack propagation, and in these many assumptions are made for the material properties, loading conditions, crack geometry, and movement. Therefore, a large number of problems rely on numerical analysis methods. Two methods are widely used.

3.3.1 Solution Based on Nodal Force Release

In the early stages, when the finite element method was applied to simulate dynamic crack propagation, crack advance was modeled by discontinued jumping. In the time increment Δt, the crack tip jumped from one node to another node of the elements along the crack direction. In order to obtain a relatively accurate solution, a rather small time increment Δt must be adopted, which was usually taken to propagate the expansion wave between the nearest two nodes of the element. In general, the crack propagation velocity is obviously lower than the wave speed; at time increment Δt, the crack actually advances only at some position between two nodes. During the process where the crack tip advances from one node to another node, the abrupt increase of crack length and sudden release of displacement restraint leads to a relatively serious high-oscillation phenomenon in the finite element solutions. To overcome this difficulty, many methods have been developed in which the force is gradually released, like the recovery force method, Lagrange multiplier method, nodal force release method, etc. Here, we concentrate on the node force release method, which is based on energy variation.

The nodal force release method removes the connecting effect of the node when the crack tip propagates to some node in the finite element mesh, which has divided the node into two nodes or multi-nodes, and then releases nodal connecting force, as shown in Figure 3.3.

In a typical generation phase computation, the input information includes the crack position as a function of time. Thus, the amount of crack advance for each time step is known. During the time stepping procedure that is used, several increments are required for the crack to traverse a single element. This differs from the extended finite element method, where the crack may propagate arbitrarily. The essence of this is necessary to design crack route and growth velocity. Since the finite element is only restrained at the corners (e.g. four-node plane elements or shell elements), the propagation of a crack along an element is simulated by the incremental release of a single nodal force. From the distance $a(t)$ or velocity $v(t)$, the time and route of the crack passing the element mesh can be estimated, and in proper order to release the nodal force. Another propagation phase method used in the analysis is that the nodal force is released by estimating the restrained nodal force reaching F_c at the crack tip location. Here, F_c is designed related to the mesh dimension and dynamic fracture toughness K_{ID}. For the given mesh, the nodal force at the crack tip is proportional to stress intensity factor K_{ID}.

In the computation of finite elements, the input includes the crack tip position as a function of time, using interpolation to obtain the distance of crack advance at each time step. During the process of crack growth along the

FIGURE 3.3 Solution of crack driving force by nodal force release method.

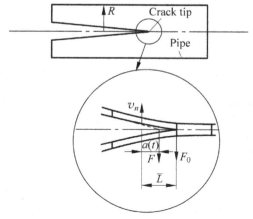

element edge, the nodal force is released step by step. At present, the energy flow to the crack tip per unit area of crack extension, G, is calculated by the finite element program. This is approximated by calculating the work done by the restraining force as follows:

$$G = \sum_{i=1}^{2} \frac{1}{h\overline{L}} \int_{\overline{i}}^{\overline{i}+\Delta t} F_i \, v_{ni} \, dt \qquad (3.27)$$

where $i = 2$ indicates that there is a pair of nodes for the two-dimensional problem, h is the material thickness, $\overline{L}$ is the length of an element along the crack advance direction, Δt is the time taken for the crack to propagate through the element, and v_{ni} is the nodal velocity normal to the direction of crack extension.

In Eq. (3.27), F is the force at the node being released, which is taken to vary with the position of the crack on the element boundary. It is expressed as follows:

$$F(t) = F_0 \left(1 - \frac{a(t)}{\overline{L}} \right)^C \qquad (3.28)$$

where F_0 is the force at the node just before node release commences and $a(t)$ is the distance that the crack has traveled along the element, as shown in Figure 3.3. The exponent C may be 1/2, 1, 3/2, 2, etc., which can be decided from the experimental data. If we take $C = 1$, it varies linearly.

3.3.2 Solution Based on Energy Balance

Another method to compute crack driving force is by using the energy balance. The basic idea in Griffith fracture theory is that there is a crack driving force, resulting from the release of potential energy in the body, driving crack growth. In addition there is resistance to crack growth associated with the necessity to supply surface energy for the newly formed crack surface. Griffith was the first to formulate this energy balance approach. This leads to a critical condition for fracture that can be written as an equality between the change in potential energy due to an incremental length of crack growth da and the resistance to this growth. This is conventionally given the symbol G in honor of Griffith. Thus, by definition we express dynamic crack propagation as

$$G = \frac{1}{h} \left(\frac{dU_e}{da} - \frac{dU_s}{da} - \frac{dU_k}{da} \right) \qquad (3.29)$$

where U_e, U_s, and U_k are the external work done on the body, internal strain energy, and kinetic energy per unit thickness respectively. h is wall thickness (crack width) and da is crack growth length. The driving force G can also be expressed by the sum of external work U_e and internal energy variation density g in the areas in front of and behind the crack:

$$Gh = \frac{dU_e}{da} + (g_s + g_k)_f - (g_s + g_k)_b \tag{3.30}$$

where g_s and g_k are strain energy density and kinetic energy density respectively. The subscripts f and b indicate the positions in front of and behind the crack tip respectively. A short length control area is usually selected, which includes the crack tip, to calculate the energy variation and analyze the values of crack driving force.

3.4 CRACK PROPAGATION IN STEADY STATE

When the maximum crack driving force that may be formed in the structure is equivalent to the material fracture toughness, there is enough energy to drive crack propagation and reach a relatively long distance. One kind of special situation for rapid crack propagation is that the crack grows at a constant speed, which is called **steady-state** crack propagation. The force and displacement variation is analyzed for the fractured structure by the solution of equations of motion.

The steady-state crack propagation problem is one where an observer traveling with a crack at a constant speed $\dot{a}$ would see no local change of the stress field during the time of observation. Although it is a dynamic problem, the load and displacement are no longer functions of time. The inertial force is the only functional coordinate. Thus, this kind of steady-state analysis is much simpler than transient analysis. As can be seen in Figure 3.4, the crack tip advances axially in the negative z-direction at constant speed $\dot{a}$ and, without loss of generality, it is assumed that the crack tip passes $z = 0$ at time $t = 0$. If a function, $\xi = z + \dot{a}t$, is introduced, the steady-state crack propagation of deformation and force are given by

$$f(x, y, z, t) = f(x, y, \xi) \tag{3.31}$$

where ξ is a moving coordinate. The crack advances at constant speed $\dot{a}$ along the ξ-direction, while the crack tip remains at the origin of the moving

FIGURE 3.4　Illustration of steady-state crack growth.

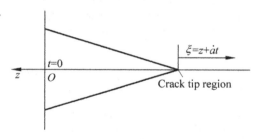

coordinate in the system. In this instance, the derivative with respect to time in equations of motion becomes the derivative with respect to coordinates:

$$\frac{\mathrm{d}}{\mathrm{d}t} = \frac{\mathrm{d}}{\mathrm{d}\xi}\frac{\mathrm{d}\xi}{\mathrm{d}t} = \dot{a}\frac{\mathrm{d}}{\mathrm{d}\xi} \quad \text{and} \quad \frac{\mathrm{d}^2}{\mathrm{d}t^2} = \dot{a}^2\frac{\mathrm{d}^2}{\mathrm{d}\xi^2} \tag{3.32}$$

Because of these exchanges, greater efficiency in solving equations of motion is achieved compared to the transient approach, in which considerable computing time is spent on crack propagation. Thus, the inertial forces can be expressed as a function of coordinate ξ. For a given structure, the variation of external work and internal energy is only considered in front of and behind the crack advancing distance Δa. Using Eq. (3.29) or (3.30), the value of the crack driving force can be calculated.

3.5 ENGINEERING APPLICATIONS OF DYNAMIC FRACTURE MECHANICS

As in many engineering applications where fracture is a concern, research and design for prevention of crack unstable propagation is an extremely important issue. This type of problem is a major research project of fracture mechanics. The purpose is to ensure that the crack driving force G (or CTOA) is less than the material fracture toughness G_d (or $(CTOA)_C$). Consequently, the cracks caused by some undetermined factors would not be rapidly propagated and the damage would be limited in the minimum local range. In this section, we use crack propagation and arrest in a pressurized gas pipeline as an example to describe an engineering application of dynamic fracture mechanics.

The general feature of a crack propagation accident is that, once initiated, the crack advances in a relatively stable velocity or arrests. It has been widely recognized that the reason for crack initiation is not completely predetermined and consequently the damage done is very serious. For instance, fractures occur in gas pipelines and in cooling pipes of nuclear reactors. For pipelines and pressurized vessels, we need to be aware that in crack unstable propagation the crack velocity is very fast. For brittle fracture, the crack velocity is about 500–800 m/s in steel pipelines. If it is in the lower temperature environment, the crack velocity is even faster, and can reach more than 1000 m/s; for ductile fracture, the crack velocity may reach 200–400 m/s for steel pipelines and 100–300 m/s for polyethylene pipelines. These statistics were computed by the US Department of Transportation on "reportable" incidents in the 560,000 km interstate gas transmission pipelines between 1970 and 1984. Of most significance is the high portion of these incidents (40%) that were caused by third-party damage (undetermined reasons). The other incidents (60%) were caused by corrosion, material defects, natural disasters, construction defects, etc. Unlike the other failures, which in principle could have been avoided by preventing initiation, third-party damage cannot be anticipated. Once initiated, the crack may rapidly propagate or quickly arrest.

The key issue in solving this problem is to compute G_{max} and determine G_d from measured experimental data, based on the criteria of crack propagation equation (3.14) and arrest equation (3.15). The classical energy balance incorporated in the novel numerical model and the restraint force release during crack tip advances are used in the computation of crack driving force, G. Schematic G data for gas pipeline fracture are shown in Figure 3.5 with decompression at low speeds and inertial effects at high speeds, resulting in a peak value of driving force at intermediate speeds. For lower speeds, the cause of driving force reduction is interior gas decompression and rapid escape behind the crack tip, resulting in a decompression limited regime. However, for higher speeds, the cause of driving force reduction is the much increased external work transformed into the inertial force of the structure, resulting in a kinetic energy increase and driving crack advancing energy being dissipated to form an inertial limited regime. The material fracture toughness G_d is slowly increased following the crack velocity increase. The energy variation in the dynamic crack propagation problem is explained in Figure 3.5.

When the damage resulting from crack propagation occurs in oil or gas transportation pipelines, whether the crack arrests depends on the one hand on the crack growth velocity and on the other on the interior decompression rate resulting from pipe leakage. For transportation liquid pipes it is quickly decompressed by the fracture, while for transportation gas pipes the decompression rate depends on the crack velocity and the gas propagation speed in the sound. The theoretical decompression effect in gas pipelines is shown in Figure 3.6 for different crack propagation velocities. The circumference arrest stress is approximately 0.3 times the crack initiation stress. It has been recognized from crack unstable propagation velocity in steel pipelines that the worse decompression effect is in brittle fracture and that a better decompression

FIGURE 3.5 Schematic prediction of crack propagation and arrest from computed dynamic crack driving force and measured dynamic crack resistance data.

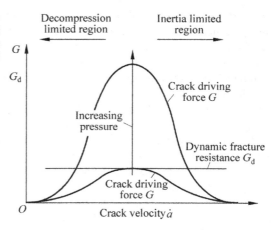

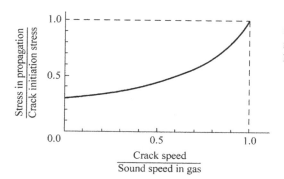

FIGURE 3.6 Schematic of theoretical decompression effect in a gas pipeline at different crack propagation velocities.

effect is in ductile fracture. Thus, it is better to choose material with good toughness in pipeline manufacture, which also has better performance in resisting stress corrosion. This is very important in the design of nuclear reactor pipes and gas pipelines. In addition, crack arrest measures are deemed to be the second line of defense against cracking accidents. Thus, the analysis and design of crack arrest members have also been paid close attention.

There are many instances where the dynamic fracture events are the result of fluid pressure acting on the structure. The crack driving force created by the fluid escaping from the fractured structure results in crack propagation. The fracture phenomenon on the fluid pressurized pipe is an interactive process involving fluid, structure, and fracture. During crack initiation and propagation in the pipeline, the internal pressure from unsteady fluid flow acts directly on the fractured pipe wall, which provides the moving force, resulting in large-scale inelastic deformation of the wall. The wall appears to undergo nonlinear unstable failure; in turn, the fluid flow is effected by the crack extension process, which produces moving boundaries on the structure. These characteristics of the fluid/structure/fracture interaction are key in developing dynamic fracture mechanics. Research on this issue is classified into computation of crack driving force G and experimental determination of material fracture toughness G_d.

Using the variation of the strain energy in the cracked pipe, an early analysis, developed by Irwin and Corten (1968), identified the crack extension force as the rate of release of circumferential strain energy stored ahead of the crack tip. Based on the fact that the pressure behind the crack tip is zero, they developed the following closed-form expression for G:

$$G_0 = \frac{\pi \, p_0^2 \, (D_0 - 2h)^2 (D_0 - h)}{8Eh^2} \tag{3.33}$$

where p_0 is the initial line pressure, D_0 and h are the outside diameter and wall thickness of the pipe, and E is Young's modulus of the material. Equating this expression to G_d yields

$$p_C = \left[\frac{8G_d E h^2}{\pi \left(D_0 - 2h\right)^2 \left(D_0 - h\right)} \right]^{1/2} \tag{3.34}$$

where the critical pressure p_C results from the Irwin–Corten analysis model.

In presenting results, it is convenient to use the dimensionless expression D_0/h for a standard dimension ratio (SDR) of a pipe. For an SDR11 polyethylene pipe with diameter 0.25 m and crack velocity 175 m/s, Figure 3.7 shows the variation of crack driving force against pressure. The value of G (gas) is calculated by the finite element code (Zhuang, 2000a) and G_0 (liquid) is calculated using Eq. (3.33). Both G and G_0 increase quickly as the initial line pressure increases, while the dimensionless values of G/G_0 remain relatively constant as the pressure increases. Thus, it is convenient to use G/G_0 and G_d/G_0 to represent the crack driving force and dynamic fracture resistance of the material in many circumstances.

Figure 3.7 shows that the value of G for gas is much higher than the value of G_0 for liquid. The latter prediction might be expected to work for a pipe pressurized by an incompressible fluid without pressure behind the crack tip. This is effectively what happens for a water pressurized pipe. For a gas pipe, however, the pressure decays steadily behind the crack tip over a characteristic length. Within this outflow region, the remaining gas can continue to develop a potent crack driving force. Pressurized water in a pipe can store as much internal energy as the pipe wall, while pressurized gas can contain a few thousand times as much because of the compressible gas. It is the internal energy that effectively determines the crack driving force. This is why the G for gas pressurization is much higher than G_0 for liquid. In turn, this is why the critical pressure for a gas pipe is much lower than that for a water pipe. This has already been proved by full-scale tests.

For rapid crack propagation in gas pipelines, a basis for this is a fluid/structure/fracture interaction model that simulates real behavior. A finite element code has been developed, called DYFRAC (You et al., 2003), which is a unitary numerical computation tool for solving complicated interaction

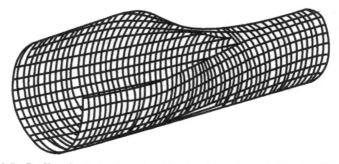

FIGURE 3.7   Profiles of calculated crack driving force G by gas and G_0 by liquid in the pipe against pressure.

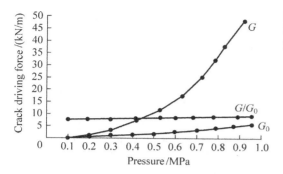

FIGURE 3.8 Demonstration of large deformation of polyethylene pipe after crack propagation.

problems. This code provides: (1) unsteady fluid flow mechanics; (2) nonlinear shell and beam structure mechanics; (3) dynamic crack mechanics; (4) a fluid/structure/fracture dynamic coupling computation model. For an SDR11 polyethylene pipe with diameter 0.25 m, Figure 3.8 demonstrates a deformation state after the crack propagation computed by the code. Following crack propagation, gas escapes behind the crack and develops a pipe wall opening, resulting in much larger deformation.

In the 1990s, following the application of gas pipelines with high strength and high toughness, for example X70 and X80 high-grade steel pipelines, the Charpy V notch (CVN) impact energy reached more than 200 J. After crack initiation, it has the characteristics of ductile crack growth and arrest. It has been proved by full-scale pipeline failure tests that although the initial crack velocity is very fast, up to 300 m/s, it decays rapidly and arrests quickly when it propagates a distance of tens to dozens of meters. This is reflected by the increase of steel toughness. The kinetic energy of driving crack propagation is dissipated and the energy release rate is rapidly reduced to lower than the material fracture toughness. Without doubt, this type of arrest phenomenon after crack propagation over a short distance without external restraint contains a material high toughness mechanism, which is good news for the gas pipeline industry.

3.6 SUMMARY

Dynamic fracture mechanics is briefly introduced in this chapter, related to the analyses of transient-state and steady-state moving crack problems and computations of dynamic stress intensity factor and energy release rate. The computations are required to solve the equations of motion, which include the moving crack structure. There are only a few analytical solutions that can be used to calculate dynamic crack growth. The solutions for large-value problems are calculated using numerical simulation methods, for example finite element programs. Nodal force release technology is described in conjunction with the conventional finite element method to simulate crack propagation. Its essence is to allocate the crack route and propagation velocity. along the element boundary. This method differs from the X-FEM, which allows arbitrary crack propagation in the element. Obviously, X-FEM has more advantages, and can represent the natural selection of crack growth. Crack propagation and arrest in pressurized gas pipelines is used as an example to explain the significance of studying dynamic fracture mechanics in engineering applications.

In Chapters 2 and 3, the essential knowledge of static and dynamic fracture mechanics has been presented in preparation for the subsequent chapters. Starting with Chapter 4, the extended finite element method will now be presented.

Fundamental Concept and Formula of X-FEM

Chapter Outline

4.1 **X-FEM Based on the Partition of Unity**   51

4.2 **Level Set Method**   53

4.3 **Enriched Shape Function**   55

  4.3.1 Description of a Strong Discontinuity Surface   55

  4.3.2 Description of a Weak Discontinuity Surface   58

4.4 **Governing Equation and Weak Form**   60

4.5 **Integration on Spatial Discontinuity Field**   64

4.6 **Time Integration and Lumped Mass Matrix**   67

4.7 **Postprocessing Demonstration**   68

4.8 **One-Dimensional X-FEM**   68

  4.8.1 Enriched Displacement   68

  4.8.2 Mass Matrix   72

4.9 **Summary**   72

4.1 X-FEM BASED ON THE PARTITION OF UNITY

Development of X-FEM is based on the partition of unity (Melenk and Babuska, 1996). The basic idea is that the conventional finite element basis can be enriched to represent a given function $\psi(x)$ on a given domain Ω in the following form:

$$\psi(x) = \sum_I N_I(x)\Phi(x) \tag{4.1}$$

where $N_I(x)$ satisfies the partition of unity, $\Sigma_I \, N_I(x) = 1$, which precisely satisfies the reproducing condition of functions (Belytschko et al., 2000). Based on this work, the undetermined parameter c_I is introduced to modify

Extended Finite Element Method. http://dx.doi.org/10.1016/B978-0-12-407717-1.00004-2

Eq. (4.1) so that the best approximation for the enriched function $\psi(x)$ can be obtained:

$$\psi(x) = \sum_I N_I(x)c_I \Phi(x) \tag{4.2}$$

In X-FEM, through adding the enriched terms based on the standard approximate displacement field, a complicated displacement field, like a discontinuous displacement field, can be described more accurately. According to this idea, the displacement field of the finite element approximation consists of two parts:

$$u^h = \sum_I N_I(x)u_I + \psi(x) \tag{4.3}$$

where $N_I(x)$ denotes the shape function of a standard finite element and u_I is the standard degree of freedom (DOF) at the finite element node. The enriched term $\psi(x)$ is used to approximate the features of the unknown displacement field. By utilizing the property of the partition of unity (Eq. (4.2)), Eq. (4.3) can be further expressed as (Belytschko and Black, 1999):

$$u^h = \sum_I N_I(x)u_I + \sum_J N_J(x)\Phi(x)q_J \tag{4.4}$$

The first term on the right-hand side is a standard finite element approximation, and the second term is an enriched term approximation based upon the partition of unity, where q_J denotes the newly added degrees of freedom at the original element node, which has no clear physical meaning and is only the undetermined coefficients used for adjusting the amplitude of the enriched function $\Phi(x)$ to get the best approximation of the actual displacement field. Equation (4.4) is the basic formulation adopted in the displacement field approximation in X-FEM. Compared with the conventional finite element, the principal difference is that additional DOFs are introduced at the element node. By introducing the definition of an enrichment shape function $\psi_J(x) = N_J(x)\Phi(x)$, Eq. (4.4) is simplified to

$$u^h = \sum_I N_I(x)u_I + \sum_J \psi_J(x)q_J \tag{4.5}$$

In Eq. (4.4), it is necessary to choose different enriched functions for different discontinuous problems. In order to increase the convergence rate, it is required that the established enriched function $\Phi(x)$ has some features of the actual displacement solution. In practical applications, $\Phi(x)$ is usually chosen from the analytical solution space, which is described in detail later. In a special situation, for example taking $\Phi(x)$ as a closed-form solution of displacement field, $u_I = 0$ and $q_J = 1$, Eq. (4.4) corresponds to the accurate expression of the actual displacement field. It is worth pointing out that the shape function N_J used for the partition of unity in the enriched term may differ from the shape function N_I used for the standard finite element term. However, the same shape function is usually used in order to facilitate solution. In Eq. (4.4), if the unknown field u^h is a vector (like the displacement field), u_I and q_J are also vectors of the same order.

It can be seen from the above description that the X-FEM is similar to p-order refinement in the conventional FEM, which does not increase the number of elements but improves the feature of the shape function to approximate the real solution. However, in p-order refinement the added nodal DOF is usually located in the interior of the element domain, while the nodal DOF added in X-FEM is still located at the original element nodes. Thus, it is convenient for solving the problem, which is a remarkable advantage of X-FEM.

The core idea of X-FEM is to use the enriched shape function basis with discontinuous features to represent the discontinuity in the element. During the computation process, the description of a discontinuous field is completely independent of the element mesh. This gives it a unique advantage in dealing with discontinuous problems, like cracks, inclusions and interfaces, etc. By constructing the enriched function $\Phi(x)$ with a special property within the element, the various discontinuous displacement fields can be captured precisely on a regular mesh. In the following sections, the two kinds of discontinuous problems will be given as examples to describe how to establish the enriched shape function.

4.2 LEVEL SET METHOD

Since X-FEM allows a discontinuous surface cross element, like a crack problem, the element mesh is independent of the fracture surface. So it is necessary to provide a geometrical description of the discontinuous surface. In X-FEM, the level set method is commonly adopted to describe the interface. On the other hand, it is usually necessary to construct the enriched shape function in X-FEM with the help of the level set function.

The level set method is one type of numerical method used for tracking discontinuity movement, like cracks, interfaces, etc. Stolarska et al. (2001) indicated that this kind of method is not necessary in X-FEM simulations. However, the level set method has quite prominent advantages: (1) the geometrical features of a discontinuity can be depicted completely by the level set function; (2) the movement of a discontinuous interface can be computed on the fixed mesh; (3) it is easily extended to the multidimensional situation.

In the level set method, a zero level set function $f(x(t),t)$ associated with the spatial and time domains is used to describe the discontinuity, which is independent of the element mesh. Because of the presence of a time variable, the level set function f is increased one more dimension than the discontinuity.

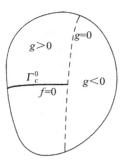

FIGURE 4.1   Level set functions f and g representing a two-dimensional crack Γ_c^0.

During the computation process, the function at the interrupted point always satisfies

$$f(x(t), t) = 0 \qquad (4.6)$$

The points that satisfy this condition in the spatial domain form a set of $\gamma(t)$. On both sides of the discontinuity, the signs for the level set function are opposite. In Figure 4.1, the location of a two-dimensional crack Γ_c^0 can be described by the zero level set function $f(x(t),t) = 0$.

One type of frequently used method for constructing the level set function is the signed distance function defined as follows (Stolarska et al., 2001):

$$f(x, t) = \pm \min_{x_\gamma \in \gamma(t)} \left\| x - x_\gamma \right\| \qquad (4.7)$$

The physical meaning of this equation is that the signed distance function at any point in the calculation domain is equal to the shortest distance from this point to the discontinuity. The points on both sides of the discontinuity have different signs. Thus, the signed distance function has the fundamental feature of the level set function: it is equal to zero at the discontinuity and has different signs on both sides of the discontinuity.

For the crack surface Γ_c^0 that terminates in the interior of the solution domain as shown in Figure 4.1, it is not enough to use only one signed distance function f because it cannot decide the position of the crack tip. For this reason, the position of a moving crack tip is described using another level set function $g(x(t),t) = 0$. Supposing the moving velocity v_i of the crack tip position x_i is known, the level set function g_i corresponding to this position can be defined as (Stolarska et al., 2001):

$$g_i = (x - x_i) \cdot \frac{v_i}{\|v_i\|} \qquad (4.8)$$

It can be seen from this equation that $g_i = 0$ represents a straight line, which passes the crack tip position x_i and is also perpendicular to velocity v_i. On both sides of the line, the signs of function g_i are opposite, which is a basic feature of the level set function.

By using this method, the position of crack surface Γ_c^0 in Figure 4.1 can be expressed by the level set function:

$$\Gamma_c^0 = \{x \in \Omega_0 | f(x) = 0 \text{ and } g(x, t) > 0\} \qquad (4.9)$$

Next, the algorithm for updating the level set functions f and g_i will be discussed. Supposing the level set functions f^n and g_i^n and the moving velocity v_i^n of crack tip x_i at time n are known, the signed distance function at time $n + 1$, f^{n+1}, is obtained:

$$f^{n+1} = f^n \quad \text{in} \quad \Omega^{\text{no update}} \qquad (4.10)$$

$$f^{n+1} = \pm \left\| (x - x_i) \times \frac{v_i}{\|v_i\|} \right\| \quad \text{in} \quad \Omega^{\text{update}} \qquad (4.11)$$

where $\Omega^{\text{no update}}$ and Ω^{update} indicate the domain of the discontinuity position with and without changes respectively. It is should be noted that the selection of positive and negative signs in Eq. (4.11) should correspond to the original signs.

In addition, the level set function g_i^{n+1} is obtained at time $n + 1$ as

$$g_i^{n+1} = g_i^n - \Delta t \|v_i^n\| \qquad (4.12)$$

The construction and update of the level set function is explained in the above content, which includes the essential factor for capturing and tracking the discontinuity in a computational domain. Such discontinuity that is irrelevant to the element mesh can be described precisely, and its evolution can be expressed strictly by the mathematical method. In the following section, we show how to connect the construction of an enriched shape function and a level set function.

4.3 ENRICHED SHAPE FUNCTION

In this section, crack and inclusion problems are used as examples to introduce the construction method of the enriched shape function; they are usually termed as "strong discontinuity" and "weak discontinuity" respectively.

4.3.1 Description of a Strong Discontinuity Surface

A finite element model for a two-dimensional plate is illustrated in Figure 4.2. A crack Γ_c^0 passes through the element mesh and terminates in the interior of the element. All of the nodes in the element mesh are marked as a set S. The nodes of crack-crossed elements are marked as a set S_h (box node in the figure). The nodes of crack-embedded elements are marked as a set S_c (circular node in the figure). The different shape functions are adopted corresponding to the node sets S_h and S_c in X-FEM.

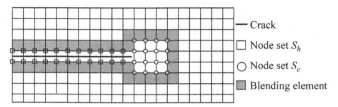

FIGURE 4.2   Enriched schemes of nodes for a crack in a 2D plate.

(1) For crack-crossed elements, the displacement fields on both sides of the crack are discontinuous, so the enriched shape function $\psi_J(x)$ may take the following form (Moës et al., 1999):

$$\psi_J(x) = N_J(x)H(f(x)) \qquad (4.13)$$

where $H(x)$ is the Heaviside step function:

$$H(x) = \begin{cases} 1 & x \geq 0 \\ -1 & x < 0 \end{cases} \qquad (4.14)$$

and $f(x)$ is the signed distance function defined as

$$f(x) = \min_{\bar{x} \in \Gamma_c^0} \|x - \bar{x}\| \cdot \text{sign}\left(\mathbf{n}^+ \cdot (x - \bar{x})\right) \qquad (4.15)$$

where $\mathbf{n}^+$ is the unit normal vector to the discontinuity line Γ_c^0. For any point x outside Γ_c^0, $f(x)$ is the shortest distance from x to Γ_c^0. The positive and negative of a distance are defined in the following way: it is positive when x lies on the same side as the direction to which $\mathbf{n}^+$ points, and it is negative when x lies on the opposite side. This type of enriched shape function is also called the Heaviside enrichment function. The Heaviside enrichment function for a four-node quadrilateral element (Q4) is illustrated in Figure 4.3. It can be found that the discontinuity occurs in the displacement field of an element because of an existing crack, which is a strong discontinuity.

(2) For crack-embedded elements, we have a node set S_c. The enriched shape function $\psi_J(x)$ may take the following form (Moës et al., 1999):

$$\psi_J(x) = N_J(x)\Phi(x) \qquad (4.16)$$

where $\Phi(x)$ could be a linear combination of the function base:

$$\Phi(x) = \left[\sqrt{r}\sin\frac{\theta}{2}, \sqrt{r}\sin\frac{\theta}{2}\sin\theta, \sqrt{r}\cos\frac{\theta}{2}, \sqrt{r}\cos\frac{\theta}{2}\sin\theta\right] \qquad (4.17)$$

In this equation, r and θ are position parameters defined in crack tip polar coordinates. It is easy to find that the above enriched function base is just the terms of the closed-form solution in a crack tip displacement field for a plane complex crack. These functions span the near crack tip asymptotic solution

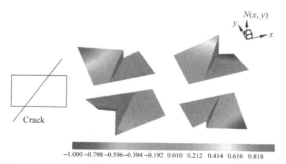

−1.000 −0.798 −0.596 −0.394 −0.192 0.010 0.212 0.414 0.616 0.818

FIGURE 4.3   Diagram of Heaviside enrichment function for a four-node quadrilateral element (Q4) crossed by a crack.

from linear elastic fracture mechanics (see section 2.2), so they are chosen not only to build the shape function discontinuity behind the crack tip but also to improve the accuracy of the captured displacement field at the crack tip. The typical enriched shape function for crack-embedded elements is called the tip enrichment function. A tip enrichment function for a four-node quadrilateral element (Q4) is illustrated in Figure 4.4. It can also be found that the discontinuity occurs in the displacement field of element because of an existing crack, which is also a strong discontinuity. During computation, it is because the idea of an enriched shape function is introduced into X-FEM that the given solution in the form of a shape function can be implemented as finite elements, for example the physical features of the solution or the obtained closed-form solution. In this way, the analytical accuracy is not only improved, but the computation cost is also greatly reduced.

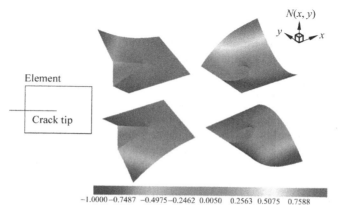

−1.0000 −0.7487 −0.4975 −0.2462 0.0050 0.2563 0.5075 0.7588

FIGURE 4.4   Diagram of tip enrichment function for a four-node quadrilateral element (Q4).

Using two types of enriched shape functions, the displacement field of a 2D cracked plate can be expressed as

$$u^h(x) = \sum_{I \in S} N_I(x)u_1 + \sum_{J \in S_h} N_j(x)H(f(x))a_J(t) + \sum_{K \in S_c} N_K(x)\Phi(x)b_K(t) \quad (4.18)$$

where a_J and b_K are enriched DOFs at the element node. The selection of node set S_c has some flexibility. It may be chosen by including one element or a few elements ahead of the crack tip, as shown in Figure 4.2. If the extended area is increased, the convergence speed may also be accelerated.

After the elements crossed by a crack or embedded by a crack tip are enriched using X-FEM, the result is a situation where only a part of the nodes have extended DOFs at the neighboring elements. The enriched shape function no longer satisfies the partition of unity in these elements. This type of element is called a **blending element**; these are illustrated as the shaded elements in Figure 4.2. These blending elements may influence computation accuracy and convergence speed.

For the Heaviside enrichment function, Belytschko et al. (2003b) proposed a modified method, which makes the shape function zero in the blending element:

$$\psi_J(x) = N_J(x)(H(f(x)) - H(f(x_J))) \quad (4.19)$$

Thus, the extended DOFs at the elements by a crossed or embedded crack no longer influence the neighboring elements. This shape function is also called a **shifted enrichment function**, and is widely used in fracture problems.

For the tip enrichment function, Fries (2008) introduced the idea of a weight function to modify the enriched shape function for the blending element:

$$\psi_J(x) = N_J(x)\Phi(x)R(x) \quad (4.20)$$

where $R(x)$ is a ramp function that decreases progressively in the blending element. This method reserves the feature of the partition of unity in the blending element. Therefore, the problem of slow convergence speed is effectively addressed, and the program implementation is facilitated.

4.3.2 Description of a Weak Discontinuity Surface

A finite element model for a 2D plate including an ellipse inclusion is shown in Figure 4.5. All of the nodes in the element mesh are marked as a set S. The nodes of the element passed by the material interface are marked as a set $S_{\text{interface}}$. Based on this example, we describe how to construct the enriched shape function for the weak discontinuity problem.

The displacement field continues at the interface between ellipse inclusion and matrix, but the derivative of the displacement field (strain field) is discontinuous, as shown in Figure 4.6. Thus, it is necessary to construct the

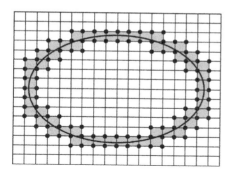

FIGURE 4.5 Finite element model for a two-dimensional plate including an ellipse inclusion and nodal enriched scheme.

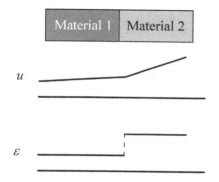

FIGURE 4.6 Distribution of displacement field and strain field at the material interface.

enriched function to satisfy this property. Using the absolute value function based on the level set, Sukumar et al. (2001) first constructed $\Phi(x)$, as given below:

$$\Phi(x) = |f(x)| \tag{4.21}$$

where $f(x)$ is level set function (Eq. (4.15)). The shape distribution of $\Phi(x)$ on the element crossed by a discontinuity is illustrated in Figure 4.6, which is also called an **interface element**. At the interface, $\Phi(x)$ indicates that the displacement u is continuous, but the derivative of displacement with respect to spatial coordinate $\varepsilon = du/dx$ (strain) is discontinuous, which satisfies the requirement of a weak discontinuity surface.

There is a shortcoming in this enriched function: it is not equal to zero at the boundary of the element containing the discontinuous interface, as shown in Figure 4.7. This means that the blending element created requires special treatment. Moës et al. (2003) proposed a modified absolute value function:

$$\Phi(x) = \sum_I |f(x_I)| N_I(x) - \left| \sum_I f(x_I) N_I(x) \right| \tag{4.22}$$

FIGURE 4.7 Shape distribution of function $\Phi(x) = |f(x)|$ on the element contained discontinuous interface.

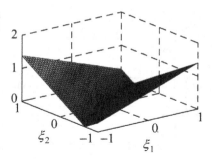

FIGURE 4.8 Shape distribution of modified absolute value function on the element contained discontinuous interface.

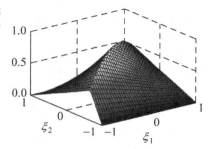

The shape distribution of the modified absolute value function on the element containing the discontinuous interface is shown in Figure 4.8. The value of Φ is zero on both sides of the element, which demonstrates that the extended DOF caused by the interface element does not influence the neighboring elements. Thus, it avoids dealing with the blending element and it is also straightforward to write such a program.

Using the modified enriched function, the displacement field for the weak discontinuous problem may be expressed as

$$u^{\mathrm{h}}(x, t) = \sum_{I \in S} N_I(x) u_I(t) + \sum_{J \in S_{\mathrm{interface}}} N_J(x) \Phi(x) q_J \qquad (4.23)$$

where q_J denotes the new added DOF at the node. $\Phi(x)$ is given in Eq. (4.22).

In the following section, the displacement field is substituted into the conventional equations of motion to obtain the finite element governing equation for X-FEM.

4.4 GOVERNING EQUATION AND WEAK FORM

Consider an initial configuration of plane domain Ω_0 with a boundary Γ_0, as shown in Figure 4.9. Γ_0^c denotes the strong discontinuity and Γ_0^{int} denotes the

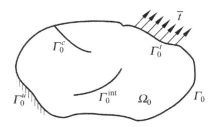

FIGURE 4.9   Initial configuration of domain Ω_0.

weak discontinuity in the domain. Conservation of momentum in terms of the nominal stress P and Lagrangian coordinates X gives

$$\frac{\partial P_{ji}}{\partial X_j} + \rho_0 b_i = \rho_0 \ddot{u}_i \quad \text{in } \Omega_0 \tag{4.24}$$

where the superposed dots denote the material time derivative, b is the body force, and ρ_0 is the material density corresponding to the initial configuration Ω_0. The kinematics equation using Green strain gives

$$E_{ij} = \frac{1}{2}\left(F_{ki} \cdot F_{kj} - \delta_{ij}\right) \tag{4.25}$$

The constitutive equation gives PK2 stress resulting in the Green strain:

$$S_{ij} = C_{ijkl} E_{kl} \tag{4.26}$$

To complete the description of the problem, the boundary conditions and initial conditions are given respectively as:

$$u_i = \bar{u}_i \quad \text{on } \Gamma_0^t \tag{4.27}$$

$$n_j^0 P_{ji} = \bar{t}_i^0 \quad \text{on } \Gamma_0^t \tag{4.28}$$

and

$$\dot{u}_i(0) = u_{i0} \quad P_{ij}(0) = P_{ij0} \tag{4.29}$$

For a strong discontinuity (crack) in the domain, there is no traction on the crack surfaces:

$$n_j^{0+} P_{ji}^+ = n_j^{0-} P_{ji}^- = 0 \quad \text{on } \Gamma_0^c \tag{4.30}$$

The contact between crack surfaces is not our concern here. Otherwise, Eq. (4.28) would be taken as a more complex form. For a weak discontinuity (interface) in the domain, there is consistent traction on the interface:

$$\|n_j P_{ji}\| = 0 \quad \text{on } \Gamma_0^{\text{int}} \tag{4.31}$$

To deduce the weak form of governing equation, the trial and test function domains are defined as follows:

$$U = \left\{ u(X,t) \,\middle|\, u(X,t) \in C^0, \; u(X,t) = \bar{u}(t) \text{ on } \Gamma_0^u, \right. \\ \left. u \text{ discontinuous on } \Gamma_0^c, \; \partial u/\partial X \text{ discontinuous on } \Gamma_0^{\text{int}} \right\} \tag{4.32}$$

$$U_0 = \left\{ \delta u(X) \,\middle|\, \delta u(X) \in C^0, \; \delta u(X) = 0 \text{ on } \Gamma_0^u, \right. \\ \left. \delta u \text{ discontinuous on } \Gamma_0^c, \; \partial(\delta u)/\partial X \text{ discontinuous on } \Gamma_0^{\text{int}} \right\} \tag{4.33}$$

where C^0 denotes a zero-order function and its derivative is only piecewise differentiable. No other difference can be found between the space of the standard FEM and the above formula, but discontinuity across a crack and derivative discontinuity across an interface are allowed here, which is the kernel idea of X-FEM.

We do not differentiate between the strong and weak discontinuity displacement fields here. For a general expression of displacement field in X-FEM as referenced in Eq. (4.5), the variational function can be expressed as

$$\delta u = \sum_I N_I(X)\delta u_I + \sum_J \psi_J(X)\delta q_J \tag{4.34}$$

We then multiply δu on both sides of momentum equation (4.24) and integrate on the initial configuration. This gives the weak form

$$\int_{\Omega_0} \delta u_i \left(\frac{\partial P_{ji}}{\partial X_j} + \rho_0 b_i - \rho_0 \ddot{u}_i \right) \mathrm{d}\Omega_0 = 0 \tag{4.35}$$

In the above equation, the nominal stress P is a function of the trail displacements. Expanding the derivative of the product in the first term and applying the fundamental theorem of calculus gives

$$\int_{\Omega_0} \delta u_i \frac{\partial P_{ji}}{\partial X_j} \mathrm{d}\Omega_0 = \int_{\Omega_0} \frac{\partial}{\partial X_j} (\delta u_i P_{ji}) \mathrm{d}\Omega_0 - \int_{\Omega_0} \frac{\partial(\delta u_i)}{\partial X_j} P_{ji} \mathrm{d}\Omega_0$$

$$= \int_{\Omega_0} \frac{\partial}{\partial X_j} (\delta u_i P_{ji}) \mathrm{d}\Omega_0 - \int_{\Omega_0} \delta F_{ij} P_{ji} \mathrm{d}\Omega_0 \tag{4.36}$$

By using Gauss theorem, the first term on the right-hand side can be converted into a contour integral:

$$\int_{\Omega_0} \frac{\partial}{\partial X_j} (\delta u_i P_{ji}) \mathrm{d}\Omega_0 = \int_{\Gamma_0} \delta u_i n_j^0 P_{ji} \mathrm{d}\Gamma_0 + \int_{\Gamma_0^c} \left(\delta u_i^+ n_j^{0+} P_{ji}^+ + \delta u_i^- n_j^{0-} P_{ji}^- \right) \mathrm{d}\Gamma_0$$

$$+ \int_{\Gamma_0^{\text{int}}} \delta u_i \left\| n_j^0 P_{ji} \right\| \mathrm{d}\Gamma_0 \tag{4.37}$$

Since $\delta u_i = 0$ on Γ_0^u, the first term on the right-hand side is reduced to an integral only on Γ_0^t. The second and the third terms both vanish due to the absence of traction on the crack surfaces of the strong discontinuity condition (4.30) and the consistent traction on the interface of the weak discontinuity condition (4.31) respectively. By combining Eqs. (4.35)–(4.37), the weak form of X-FEM is obtained to find $u \in U$, such that

$$\int_{\Omega_0} \left(\delta F_{ij} P_{ji} - \delta u_i \rho_0 b_i + \delta u_i \rho_0 \ddot{u}_i \right) d\Omega_0 - \int_{\Gamma_0^t} \delta u_i \bar{t}_{i0} d\Gamma_0 = 0 \qquad \forall \, \delta u \in U_0 \quad (4.38)$$

This looks quite similar to the weak form of standard FEM while the trial and test functions are changed here. The expression is simplified because of the assumption that no contact and friction between the crack surfaces exists. The weak form will be appended with more complicated terms if contact and friction are involved. By substituting the variational displacement field equation (4.34) into Eq. (4.38), the weak form of momentum equation (4.24) can be obtained.

The internal node force is given by

$$f_{iI}^{\text{int}} = \int_{\Omega_0} \left(B_{Ij} \right)^{\text{T}} P_{ji} d\Omega_0 \tag{4.39}$$

In the above, the matrix B can be written as a Voigt form:

$$B = \left[B^{\text{fem}}, B^{\text{enr}} \right] \tag{4.40}$$

where

$$B_I^{\text{fem}} = \begin{bmatrix} N_{I,x} & 0 \\ 0 & N_{I,y} \\ N_{I,y} & N_{I,x} \end{bmatrix}, \quad B_J^{\text{enr}} = \begin{bmatrix} \psi_{J,x} & 0 \\ 0 & \psi_{J,y} \\ \psi_{J,y} & \psi_{J,x} \end{bmatrix}$$

The internal node force may be linear and the stiffness matrix is given by

$$K = K^{\text{mat}} + K^{\text{geo}} \tag{4.41}$$

where the material stiffness and geometrical stiffness matrix are respectively expressed as

$$K_{IJ}^{\text{mat}} = \int_{\Omega_0} B_I^{\text{T}} \left[C^{\text{SE}} \right] B_J d\Omega_0, \quad K_{IJ}^{\text{geo}} = I \int_{\Omega_0} B_I^T S B_J d\Omega_0 \tag{4.42}$$

The discretized momentum equation set may be written in matrix form, such that

$$\begin{bmatrix} M_{uu} & M_{uq} \\ M_{uq}^{\text{T}} & M_{qq} \end{bmatrix} \begin{pmatrix} \ddot{u} \\ \ddot{q} \end{pmatrix} + \begin{bmatrix} K_{uu} & K_{uq} \\ K_{uq}^{\text{T}} & K_{qq} \end{bmatrix} \begin{pmatrix} u \\ q \end{pmatrix} = \begin{pmatrix} f^{\text{ext}} \\ Q^{\text{ext}} \end{pmatrix} \tag{4.43}$$

This is an equation set of X-FEM, which should be solved to get the final results. Here, the solving variables, except for the original DOF u, include the added DOF q to reflect the internal discontinuity. K_{ua} is a coupling term from two kinds of DOFs in the stiffness matrix, which can be solved using Eq. (4.42). f^{ext} and Q^{ext} are external forces corresponding to node DOFs respectively, as given below:

$$f_{il}^{\text{ext}} = \int_{\Gamma_0^t} N_I \bar{t}_{i0} d\Gamma_0, \qquad Q_{il}^{\text{ext}} = \int_{\Gamma_0^t} \psi_I \bar{t}_{i0} d\Gamma_0 \tag{4.44}$$

The terms in the mass matrix can be expressed as

$$M_{IJ}^{uu} = \int_{\Omega_0} \rho_0 N_I N_J d\Omega_0, \quad M_{IJ}^{qq} = \int_{\Omega_0} \rho_0 \psi_I \psi_J d\Omega_0, \quad M_{IJ}^{uq} = \int_{\Omega_0} \rho_0 N_I \psi_J d\Omega_0$$

$$\tag{4.45}$$

It can be found that there also are coupling terms in the mass matrix. In practice programming, the same mass matrix is adopted as that in the conventional FEM. It is necessary to ignore the coupling terms, so only the diagonalization mass matrices M_{uu} and M_{qq} are used. The diagonalization method is introduced in section 4.6.

For the weak form expression, there are very similar expressions for node force and mass matrix corresponding to X-FEM as compared with conventional FEM, although a redundant DOF is introduced in X-FEM. The difference lies in employing a different shape function ψ. Thus, it is easy to add X-FEM capability based on the conventional FEM program.

4.5 INTEGRATION ON SPATIAL DISCONTINUITY FIELD

Because the feature of the shape function has been changed in X-FEM, the shape function is no longer constructed by the smooth polynomial in conventional FEM. This presents a challenge in carrying out precise integration for the element internal force and mass matrix. For example, in a fracture problem, both integrands in Eqs. (4.42) and (4.45) are no longer continuous functions. Large errors will be introduced if the Gauss integral scheme is still used in conventional FEM. In the following, two methods are described to deal with the integral for the discontinuity function, which are frequently adopted in X-FEM computations (Belytschko et al., 2001):

1. Integral point density is uniformly increased. For example, although some errors may be found in a plane 5×5 integral, these errors are allowable in engineering applications.
2. A subdomain integral method is used for precise integration of the discontinuity field. A one-dimensional element is taken as an example to

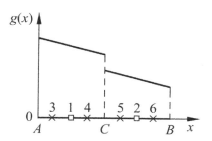

FIGURE 4.10   Integration diagram for discontinuity function $g(x)$ in a one-dimensional element subdomain.

explain this method. An integration diagram for the discontinuity function $g(x)$ in element AB is illustrated in Figure 4.10. If the traditional integral scheme is adopted, Gaussian points 1 and 2 are selected, and large errors can be introduced. In order to reach the goal of precise integration, the element AB can be divided into an element AC and an element CB corresponding to the discontinuity points. Because the function $g(x)$ is separately continued at AC and CB, Gaussian integration points 3 and 4, as well as 5 and 6, are chosen corresponding to elements AC and CB to carry out precise integration. Superposing the two parts of these integral results, the integration of discontinuity function $g(x)$ is obtained at the element AB. It is necessary to indicate that this process does not increase new DOFs at node C. The values at integral points 3, 4, 5, and 6 are obtained by interpolation from the results at nodes A and B.

Figures 4.11 and 4.12 are schematic diagrams of integration subdomains in the two-dimensional and three-dimensional situations respectively. The elements including discontinuity are divided by trianglular or tetrahedral subdomains. The integrals are carried out corresponding to different subdomains, and then the results are superposed. Strouboulis et al. (2000) proposed another subdomain divided method, but the idea is very similar.

We do not adopt either of the two methods in our program, because both have some shortcomings. The biggest weakness of the first method is the very

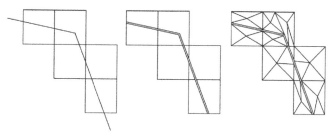

FIGURE 4.11   Schematic diagrams of integration subdomain in a two-dimensional situation (Belytschko et al., 2001).

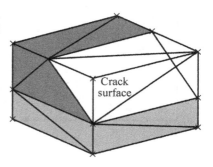

low precision. The element divided by a discontinuity has a very small area, and clearly the 5×5 integration may not be precise for the stiffness matrix in such a small area. It is found that a distinct error may be introduced in this case. If precise integration is expected on both areas of discontinuity, the second method may obviously accomplish this. However, the repartition subdomains may increase the computational cost. Therefore, the integral scheme applied in reality may be a compromise between losing precision and increasing computational cost. The integral scheme that we used is illustrated in Figure 4.13.

For each element with discontinuity, the minimum length divided by the discontinuity along the X and Y directions is first solved, which is called the **characteristic length**. The principle of assigned integral points is that they are uniformly distributed in the element, and it is guaranteed that there are at least two points existing in this characteristic length, as shown in Figure 4.13. There are 6×7 integration points in this figure. For different elements, the distribution of integral points must be calculated separately. From our experience

FIGURE 4.13 Integral scheme.

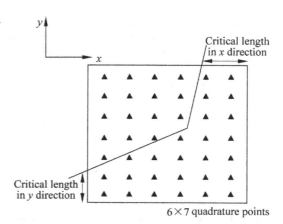

6×7 quadrature points

we have observed that this method guarantees a certain precision, while its computational cost is less than the subdomain integral method.

4.6 TIME INTEGRATION AND LUMPED MASS MATRIX

In dynamic fracture problems, explicit time integration is usually adopted for the solution of momentum equation (4.43):

$$u_{t+\Delta t} = u_t + \Delta t \dot{u}_t + \frac{1}{2}\Delta t^2 \ddot{u}_t \tag{4.46}$$

$$\dot{u}_{t+\Delta t} = \dot{u}_t + \frac{1}{2}\Delta t(\ddot{u}_t + \ddot{u}_{t+\Delta t}) \tag{4.47}$$

$$\ddot{u}_{t+\Delta t} = M^{-1}\left(f_{t+\Delta t}^{\text{ext}} - f_{t+\Delta t}^{\text{int}}\right) \tag{4.48}$$

where u_t, $\dot{u}_t$, and $\ddot{u}_t$ are displacement, velocity, and acceleration at time t respectively. Δt is time increment, f^{ext} and f^{int} are node external force and internal force respectively, and M is the mass matrix.

The similarity with conventional FEM is that the lumped mass matrix is used in the program to improve solution efficiency. The difference with conventional FEM is that the element is divided by discontinuity. If there is a crack, it means the materials on both sides are separated; if there is an interface, it means the materials on both sides have different material properties. As a result, the mass in each element is not equally assigned to node DOFs in X-FEM, which is unlike conventional FEM, which uniformly distributes the mass to each node. At present, there is no maturity theory on diagonalization of the mass matrix in X-FEM. Menouillard et al. (2006) made a systematic comparison of the influence of calculation precision and stable time step for a few types of diagonal methods that are often used, like the row superposition method and the mass equally assigned method.

In our programming approach (Zhuang and Cheng, 2011a, b, c), both sides of the mass are calculated separately corresponding to the element crossed by the crack, and then each side of the mass is reassigned by averaging to the corresponding nodes. Since it is necessary to find the area of an irregular polygon in the mass calculation, the computational cost may be increased. Therefore, on the premise of guaranteeing precision, an approximate and convenient method is used in the program. As mentioned previously, the integral points are distributed uniformly in the element, so the mass m_i at node i can be expressed as

$$m_i = \left(\frac{n_{\text{int}}}{N_{\text{int}}} \cdot m_{\text{ele}}\right) \Big/ s \tag{4.49}$$

(a)   **(b)**

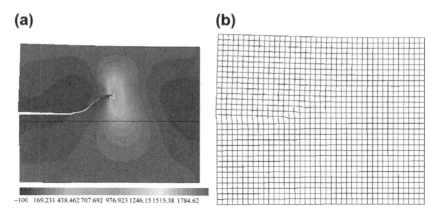

-100 169.231 438.462 707.692 976.923 1246.15 1515.38 1784.62

FIGURE 4.14   Postprocessing demonstration of a crack.

where N_{int} is the number of integral points in the element, while n_{int} is the number of integral points in the subdomain of an element that only includes node i, m_{ele} is element mass, and s is the number of nodes in the subdomain of an element that only includes node i.

Equation (4.49) is suitable for the crack situation, but it needs to be modified for the interface situation. Since there are different material properties on both sides of the interface, it is necessary to deal with the mass individually. However, the similar mass assigned approach may be used, which is introduced in detail in Chapter 8.

4.7 POSTPROCESSING DEMONSTRATION

In the computation of X-FEM, the element includes discontinuity. An arbitrary shape discontinuity can be captured in a regular mesh, as shown in Figure 4.14(b). It is expected to be clear immediately where the crack is located in the output results. Thus, similar to subdomain integration, it is necessary to divide the element into subdomains, as shown in Figure 4.14(a). The segmented subdomains are only used to demonstrate postprocessing and are not involved in the computation. This is an essential difference compared with the conventional FEM.

4.8 ONE-DIMENSIONAL X-FEM

In this section, we consider fracture in one dimension, which is simple but representative, as an example to further illustrate the basic concept of X-FEM.

4.8.1 Enriched Displacement

The crack is at one point in one dimension and the structure is divided into two parts by this crack. As shown in Figure 4.15, the one-dimensional element

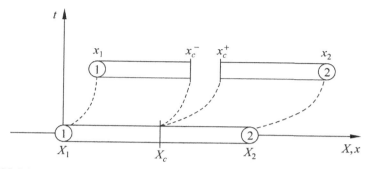

FIGURE 4.15   1D cracked element in initial configuration and current configuration.

with two nodes at X_1 and X_2 is cut by a crack at X_c. The element is originally of length l_0 and constant cross-sectional area A_0. We assume there is no jump across the crack initially, as shown in the lower part of Figure 4.15. After a time period t, the two end nodes have displacements u_1 and u_2, and the two crack surfaces are separated with displacements u_c^- and u_c^+, as shown in the upper part of Figure 4.15. Next we will model the displacement field of the element with conventional FEM and X-FEM.

First, the displacement field is modeled using the conventional FEM. Because the FEM is incapable of describing a discontinuous field, the element must be divided into two independent elements and another two nodes are defined at the crack position, as shown in Figure 4.16.

Linear interpolation is adopted and the displacement field can be described as

$$u(X,t) = \begin{cases} \dfrac{X_c - X}{X_c - X_1} \cdot u_1 + \dfrac{X - X_1}{X_c - X_1} \cdot u_c^-, & X_1 \leq X \leq X_c \quad (a) \\[3mm] \dfrac{X_2 - X}{X_2 - X_c} \cdot u_c^+ + \dfrac{X - X_c}{X_2 - X_c} \cdot u_2, & X_c < X \leq X_2 \quad (b) \end{cases} \quad (4.50)$$

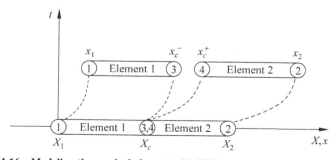

FIGURE 4.16   Modeling the cracked element with FEM.

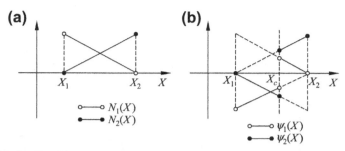

FIGURE 4.17   Heaviside enriched shape functions.

In contrast, using X-FEM we can model the discontinuous field with just the two endpoints being nodes. Each endpoint has two degrees of freedom, u_i^H and q_i^H, $i = 1, 2$, and the Heaviside enrichment function in Eq. (4.13) is adopted. In one dimension, the enriched function can be written as

$$\psi_J(X) = N_J(X)H(X - X_c) \tag{4.51}$$

in which the signed distance function has been simplified as $f(X) = X - X_c$. The Heaviside enriched shape functions are plotted in Figure 4.17.

The displacement field given with the enriched terms is

$$u(X, t) = \frac{1}{l_0} \left\{ \begin{array}{l} X_2 - X \\ X - X_1 \\ (X_2 - X)H(X - X_c) \\ (X - X_1)H(X - X_c) \end{array} \right\}^{\mathrm{T}} \left\{ \begin{array}{l} u_1^H(t) \\ u_2^H(t) \\ q_1^H(t) \\ q_2^H(t) \end{array} \right\} \tag{4.52}$$

To demonstrate that the field described by Eq. (4.52) is the same as that by Eq. (4.50), we express $\{u_1, u_2, u_c^-, u_c^+\}$ with $\{u_1^H, u_2^H, q_1^H, q_2^H\}$:

$$u_1 = u_1^H - q_1^H$$

$$u_2 = u_2^H + q_2^H$$

$$u_c^- = \frac{1}{l_0} \left[(X_2 - X_c)u_1^H + (X_c - X_1)u_2^H - (X_2 - X_c)q_1^H - (X_c - X_1)q_2^H \right]$$

$$u_c^+ = \frac{1}{l_0} \left[(X_2 - X_c)u_1^H + (X_c - X_1)u_2^H + (X_2 - X_c)q_1^H + (X_c - X_1)q_2^H \right]$$

$$\tag{4.53}$$

For the left part of the structure divided by the crack, the displacement field is expressed by Eq. (4.50a). By substituting $\{u_1, u_2, u_c^-, u_c^+\}$ in Eq. (4.53) into Eq. (4.50a), we get

$$u(X,t)_{\text{left}} = \frac{1}{X_c - X_1} \begin{Bmatrix} X_c - X \\ X - X_1 \end{Bmatrix}$$
$$\begin{Bmatrix} u_1^H - q_1^H \\ \frac{1}{l_0} [(X_2 - X_c)(u_1^H - q_1^H) + (X_c - X_1)(u_2^H - q_2^H)] \end{Bmatrix} \quad (4.54)$$

Substituting $l_0 = X_2 - X_1$ into the above equation, the expression can be simplified to

$$u(X,t)_{\text{left}} = \frac{1}{l_0} [(X_2 - X)(u_1^H - q_1^H) + (X - X_1)(u_2^H - q_2^H)] \quad (4.55)$$

As for the X-FEM expression, Eq. (4.52), the left part of the element has the value $H(X - X_c) = -1$ and then the displacement field can be written as

$$u(X,t)_{\text{left}} = \frac{1}{l_0} \begin{Bmatrix} X_2 - X \\ X - X_1 \\ -(X_2 - X) \\ -(X - X_1) \end{Bmatrix}^T \begin{Bmatrix} u_1^H(t) \\ u_2^H(t) \\ q_1^H(t) \\ q_2^H(t) \end{Bmatrix} \quad (4.56)$$

which is the same as Eq. (4.55). So the two fields modeled by FEM and X-FEM are equivalent.

We can also model the X-FEM element with shifted Heaviside function as in Eq. (4.19) for the blending elements near to the discontinuity element in order to satisfy the partition of unity. In one dimension, the shifted Heaviside function can be written as:

$$\psi_J(x) = N_J(x)(H(X - X_c) - H(X_J - X_c)) \quad (4.57)$$

which is plotted in Figure 4.18. Readers can write the interpolation equations using the same method as mentioned above.

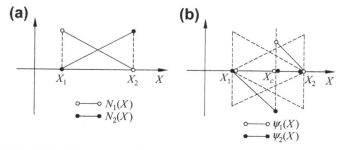

FIGURE 4.18   Shifted Heaviside enrichment shape functions.

4.8.2 Mass Matrix

The expression of the mass matrix of X-FEM elements is given in Eq. (4.45). For the one-dimensional element depicted in Figure 4.15, the initial density is ρ_0 and we define the parameter $c = (X_c - X_1)/l_0$. Then the consistent mass matrix can be written as

$$
M = \rho_0 l_0 A_0 \left\{
\begin{array}{cccc}
\dfrac{1}{3} & \dfrac{1}{6} & \dfrac{1}{3} - 2c + 2c^2 - \dfrac{2}{3}c^3 & \dfrac{1}{6} - c^2 + \dfrac{2}{3}c^3 \\[2mm]
\dfrac{1}{6} & \dfrac{1}{3} & \dfrac{1}{6} - c^2 + \dfrac{2}{3}c^3 & \dfrac{1}{3} - \dfrac{2}{3}c^3 \\[2mm]
\dfrac{1}{3} - 2c + 2c^2 - \dfrac{2}{3}c^3 & \dfrac{1}{6} - c^2 + \dfrac{2}{3}c^3 & \dfrac{1}{3} & \dfrac{1}{6} \\[2mm]
\dfrac{1}{6} - c^2 + \dfrac{2}{3}c^3 & \dfrac{1}{3} - \dfrac{2}{3}c^3 & \dfrac{1}{6} & \dfrac{1}{3}
\end{array}
\right\}
$$

(4.58)

According to the diagonalization method introduced in Chapter 4.6, the lumped mass matrix of the one-dimensional element is

$$
M_{\text{lumped}} = \rho_0 l_0 A_0
\begin{bmatrix}
\dfrac{1}{2} & 0 & 0 & 0 \\[2mm]
0 & \dfrac{1}{2} & 0 & 0 \\[2mm]
0 & 0 & c & 0 \\[2mm]
0 & 0 & 0 & 1 - c
\end{bmatrix}
$$

(4.59)

4.9 SUMMARY

The basic idea of X-FEM is to use an enriched shape function to express the discontinuity. The feature of discontinuity is implicitly included in the construction information of the shape function. The position and configuration of the discontinuity does not depend on the element mesh. Because the enriched shape function is enhanced, the extended DOF is added on the original node, which results in an increase of nodal external force, mass matrix, and stiffness matrix respectively in the solution scheme.

The essential concept and formula of X-FEM, as well as the technical process of program implementation, have been discussed in this chapter. The partition of unity and level set method are described. The expressions for enriched shape functions are provided, which are in terms of the strong form including crack-crossed and crack-embedded elements. The enriched shape functions are also given for the strain discontinuous interface, in terms of the weak form. The discretized momentum equation of X-FEM deduced from the weak form is obtained. Schemes of time and spatial integration and lumped mass matrix distribution are also provided. Fracture in one dimension is taken as an example to illustrate the basic concept of X-FEM.

In Chapter 5, some numerical examples in 2D are provided in detail to further describe the applications of X-FEM and the methods used to deal with the programming problems.

Numerical Study of Two-Dimensional Fracture Problems with X-FEM

Chapter Outline

5.1 **Numerical Study and Precision Analysis of X-FEM**   76
   5.1.1 A Half Static Crack in a Finite Plate   76
   5.1.2 A Beam with Stationary Crack under Dynamic Loading   77
   5.1.3 Simulation of Complex Crack Propagation   77
   5.1.4 Simulation of the Interface   79
   5.1.5 Interaction Between Crack and Holes   81
   5.1.6 Interfacial Crack Growth in Bimaterials   83

5.2 **Two-Dimensional High-Order X-FEM**   84
   5.2.1 Spectral Element-Based X-FEM   84
   5.2.2 Mixed-Mode Static Crack   87
   5.2.3 Kalthoff's Experiment   89
   5.2.4 Mode I Moving Crack   91

5.3 **Crack Branching Simulation**   93
   5.3.1 Crack Branching Enrichment   94
   5.3.2 Branch Criteria   95
   5.3.3 Numerical Examples   97
     (a) Static Example   97
     (b) Dynamic Branching   99

5.4 **Summary**   101

In this chapter, some examples are given in section 5.1 to verify the X-FEM program; these include calculating the stress intensity factor for steady-state cracks, complex crack growth, interaction between crack and cavity etc., as well as simulating other two-dimensional discontinuity problems. The precision of simulation results is discussed, and then the reliability and robustness are verified for the X-FEM program developed by the authors and colleagues.

Extended Finite Element Method. http://dx.doi.org/10.1016/B978-0-12-407717-1.00005-4

Some numerical results are compared with the analytical solutions to prove the capability and efficiency of the algorithm and program. In section 5.2, two-dimensional high-order X-FEM is developed based on the spectral element method, which can be used to control the numerical disturbance occurring in simulating dynamic fracture using low-order elements. It is very efficient in improving the simulation precision for rapid crack propagation. For the calculation of stress intensity factor in these examples, the interaction integral method is used, which has been given in Chapters 2 and 3 corresponding to quasi-static fracture and dynamic fracture. In addition, the fracture criterion of the maximum circumference tension stress is also adopted. In section 5.3 the standard enrichment function for a single crack in one element is modified to deal with branched cracks. We show two kinds of new enriched elements, namely the element containing two cracks and the element with a junction. Another series of enriched degrees of freedom is introduced to provide the additional discontinuous characteristics in the element. A shifted enrichment scheme is used. The branching process of a single mode I crack is simulated, and the branching angle and propagation path are given.

5.1 NUMERICAL STUDY AND PRECISION ANALYSIS OF X-FEM

5.1.1 A Half Static Crack in a Finite Plate

This example is used to test computing accuracy for a static crack. A half crack is located on the left side of a finite rectangular plate with a 41×41 element mesh, as shown in Figure 5.1(a). The tension force acts uniformly on the top and bottom boundaries. Changing crack length, we can get a couple of

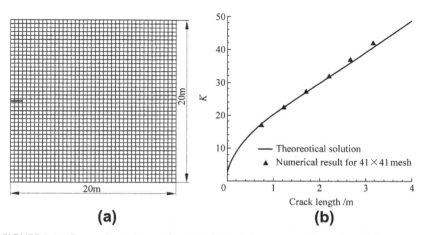

(a)　　　　　　　　　　　　**(b)**

FIGURE 5.1　Comparison of stress intensity factor between X-FEM results and theoretical solutions. (a) Element mesh. (b) Stress intensity factor versus crack length.

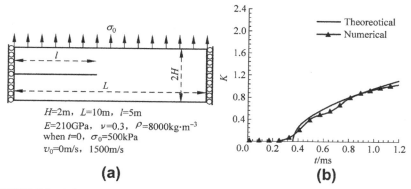

H=2m, L=10m, l=5m
E=210GPa, ν=0.3, P=8000kg·m^{-3}
when t=0, σ_0=500kPa
v_0=0m/s, 1500m/s

(a)   **(b)**

FIGURE 5.2   A beam with a stationary crack under dynamic loading.

values of stress intensity factor (SIF), which are compared with the theoretical solutions. The results agree very well, as shown in Figure 5.1(b).

5.1.2 A Beam with Stationary Crack under Dynamic Loading

A beam with stationary crack under dynamic loading is illustrated in Figure 5.2(a). The parameters are also listed below the beam. It is loaded on the upper edge by uniform tension at time $t = 0$. The SIF remains at zero until 0.38 ms approximately, when the dilation wave arrives at the crack tip. Freund (1990) has proposed a theoretical solution, which is plotted in Figure 5.2(b), together with the numerical results using X-FEM. Acceptable agreement is found between the two.

5.1.3 Simulation of Complex Crack Propagation

A simply supported beam, as shown in Figure 5.3, has an initial notch located at $d = c \cdot L/2$. An impact load is acting on the top of the beam. Since this load

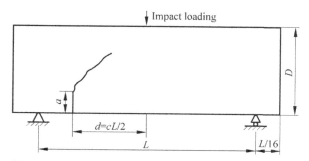

FIGURE 5.3   Complex crack growth in a simple supported beam under impact load.

is in the middle location departure from the notch vertical line, it must generate a complex crack at the notch tip, which propagates alone a zigzag route and toward the loading position. The relevant geometry and material parameters are provided by: $L = 203.2$ mm, $D = 76.2$ mm, $a = 19.05$ mm, thickness $B = 25.4$ mm, density $\rho = 2400$ kg/m^3, Young's modulus $E = 31.37$ GPa, Poisson's ratio $\nu = 0.2$. John (1990) carried out experiments with specimens of the same dimensions. The measured data show that the included angle between crack growth route and vertical orientation is equal to 22° at $c = 0.5$. Some simulations have been carried out by Belytschko and Tabbara (1996) using a meshless method, and Song et al. (2006) using X-FEM.

In the computation here, the impact load is applied in terms of velocity and adopted in the same quantity as given by Song et al. (2006):

$$v_0 = \begin{cases} 0.065m/t_1 & t \leq t_1 \\ 0.065m/s & t > t_1 \end{cases} \qquad (5.1)$$

where $t_1 = 196$ μs. The contour plot of computed Mises stress distribution is given in Figure 5.4 during the crack growth process. Crack initiation occurs at 647 μs and stopping propagation at 720 μs. During this period, the crack

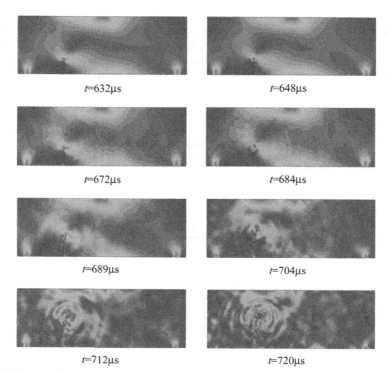

$t=632$μs $t=648$μs

$t=672$μs $t=684$μs

$t=689$μs $t=704$μs

$t=712$μs $t=720$μs

FIGURE 5.4   Contour plots of computed Mises stress (Zhuang and Cheng, 2011a).

FIGURE 5.5   Average angle between measured crack growth route and vertical direction.

propagation distance is about 28.5 mm. Thus, the average velocity of crack growth is equal to 390 m/s.

After crack initiation, the crack propagates in the direction of the concentration loading point, which demonstrates the features of complex cracks. Regular square meshes are arranged in the computation, while the crack growth route appears in a zigzag shape, which does not depend on the mesh assignment. This is a unique advantage of X-FEM, which can precisely capture the actual crack growth configuration. The average angle between the measured crack growth route and the vertical direction is α, as shown in Figure 5.5. The computed angle is about 21° by X-FEM, which agrees well with the value of 22° measured experimentally.

5.1.4 Simulation of the Interface

The reliability of crack growth simulation used by the X-FEM program, which has been developed by the authors and colleagues, has been tested and verified in the previous examples. The veracity of the weak discontinuity problem simulation is examined in this subsection. The shape function constructed in Eq. (4.22) is used, which has the features of the discontinued strain field at the interface and a continuous displacement field in the domain. This is called weak discontinuity in the element, for example inclusion, bimaterial interface, etc. This is coincident with the deformation at a two-phase material interface; thus, it can be used to model the interface independently of the mesh.

The contour plot of stress simulated by X-FEM is shown in Figure 5.6(a) for bimaterial under uniform tension acting at the top and bottom boundaries. The black rectangular mesh expresses the element and the interface is also indicated in the figure. It is seen that the interface may pass through the element mesh, which results in the strain along the vertical direction being discontinued on both sides of interface. This can be proved by the stress balance condition and the simple relationship between stress and strain on both sides of the interface.

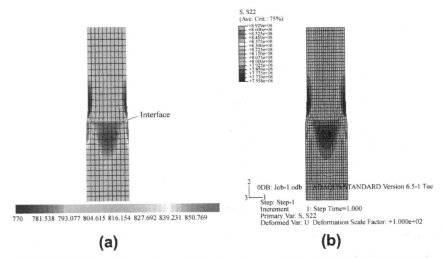

(a)   **(b)**

FIGURE 5.6 Deformation contour plot for two-phase material with interface (output: σ_2). (a) X-FEM result. (b) FEM result (ABAQUS).

The same problem can be addressed using the commercial FEM code ABAQUS. Using the conventional FEM and double increased mesh density, the results obtained are plotted in Figure 5.6(b). When adopting the conventional FEM, the interface is coincident with the element mesh. However, the tendency of integral deformation is consistent with the results computed using X-FEM. For further comparison, the vertical stresses σ_2 calculated by X-FEM and FEM respectively are output along a center axial line. The comparison results are illustrated in Figure 5.7, and are essentially consistent.

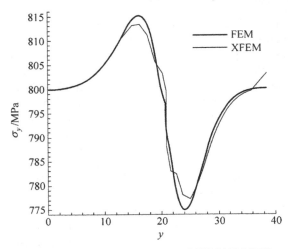

FIGURE 5.7 Comparison results between X-FEM and FEM (ABAQUS).

5.1.5 Interaction Between Crack and Holes

Taking advantage of X-FEM, the interactions among the crack, inclusion, hole, etc. can be conveniently simulated in regular meshes. The plate is fixed on the right-side edge and loaded by uniform tension on the upper and lower boundaries, as shown in Figure 5.8. Holes ahead of the crack tip have two different forms: one aligned with the crack tip (Figure 5.8(a)) and the other deviating a small distance from the crack tip (Figure 5.8(b)).

The value of the stress intensity factor (SIF) for a crack without a hole is used for comparison: $K_I = 911$ MPa·m$^{1/2}$. Due to the existence of the hole, the SIF for mode I only is augmented to $K_I = 1036.1$ MPa·m$^{1/2}$ in the left-side situation, and the SIFs for mixed mode I and mode II are augmented to $K_I = 950.3$ MPa·m$^{1/2}$ and $K_{II} = 92.4$ MPa·m$^{1/2}$ in the right-side situation. Also, there is a growth angle with $-10.9°$ deviation from the original direction, which suggests that the crack is attracted by the hole.

The element crossed by a crack and hole is shown in Figure 5.8. It is independent of the mesh because of the advantages of X-FEM mentioned previously. Is it accurate enough? The following two examples will answer this.

The problem of element-embedded cracks and holes in an infinite plane was solved analytically by Hu et al. (1993). Their results are used as a benchmark here. Figures 5.9 and 5.11 present the configuration and mesh for the hole aligned with the crack tip and deviating a small distance from the crack tip respectively. Figures 5.10 and 5.12 show the comparisons between the analytical solutions of Hu et al. (1993) and the numerical results obtained by X-FEM corresponding to the situations shown in Figures 5.9 and 5.11. The SIFs are normalized with respect to that of a single crack in an infinite plane. There is a strong interaction effect for SIF values when the crack tip is very

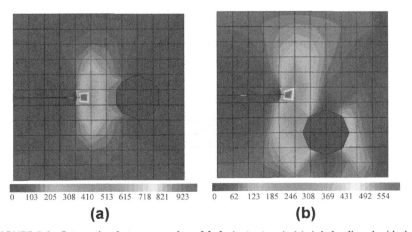

| 0 | 103 | 205 | 308 | 410 | 513 | 615 | 718 | 821 | 923 |

(a)

| 0 | 62 | 123 | 185 | 246 | 308 | 369 | 431 | 492 | 554 |

(b)

FIGURE 5.8 Interaction between crack and hole (output: σ_2). (a) A hole aligned with the crack tip. (b) A hole deviating a small distance from the crack tip.

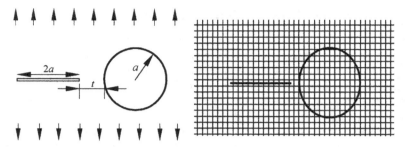

FIGURE 5.9 Configuration and mesh for a hole aligned with the crack tip.

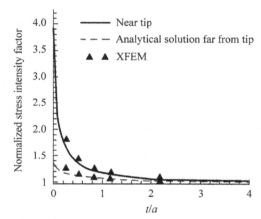

FIGURE 5.10 Interaction results for a hole aligned with the crack tip.

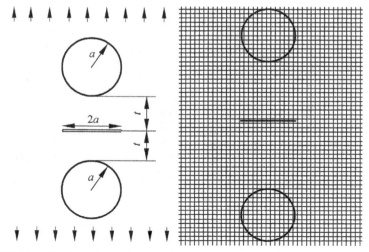

FIGURE 5.11 Configuration and mesh for holes deviating a small distance from the crack tip.

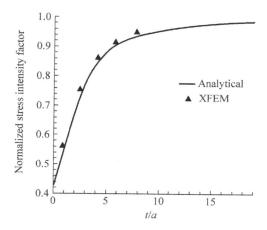

FIGURE 5.12 Interaction results for holes deviating a small distance from the crack tip.

close to the hole. Both examples show reliable accuracy of the X-FEM method and the program that we have developed. The relative error is less than 5% for the majority of the curve. It is around 10% when the crack lies very close to the hole.

5.1.6 Interfacial Crack Growth in Bimaterials

A specimen fixed on the right-side edge is loaded by uniform tension on both the upper and lower surfaces. The initial crack is located on the geometric axis and propagation starts from the left side. This would be a typical mode I crack if the material was homogeneous. However, the specimen consists of two kinds of materials: one is steel with $E_s = 210$ GPa, $K_{Ics} = 150$ MPa·m$^{1/2}$, and the other is aluminum with $E_s = 69$ GPa, $K_{Ics} = 37$ MPa·m$^{1/2}$. The interface between the two materials is below the crack by $h/20$, where h is the height of the specimen. There are two types of test: the first test is aluminum on the upper part and steel on the lower part; the second test is steel on the upper part and aluminum on the lower part. We can obtain the configurations of crack propagation.

The computed results illustrate that the crack tends to deviate from a straight line when material heterogeneousness is introduced. Whether steel is on the upper or lower part, the crack always propagates into aluminum, which is the softer part. This effect can be quantified. Steel is in the lower part while the upper part is no longer aluminum but has a material modified Young's modulus. The ratio of the Young's modulus (E_{up}/E_{low}) can be recorded and plotted versus the initial propagating angle of the crack. The angle is zero when $E_{up}/E_{low} = 1$, which is homogeneous material. It is positive for an ascending crack when $E_{up}/E_{low} < 1$ and it is negative for a descending crack when $E_{up}/E_{low} > 1$. The angle descends monotonously with increasing ratios of Young's modulus. So we conclude that the crack always tends to propagate into the softer medium and the propagating angle increases if the material difference is greater.

5.2 TWO-DIMENSIONAL HIGH-ORDER X-FEM

In the preceding sections, all the shape functions in X-FEM are constructed by linear functions, called low-order finite elements. One difficulty with low-order finite elements, both in the context of X-FEM and other applications to dynamic problems, is that they sometimes do not have the desired accuracy. They exhibit rather marked wave oscillations. The numerical oscillations are strongly amplified during dynamic crack propagation (Menouillard et al., 2010), and further affect the calculation of crack propagation direction and speed. In this section a high-order X-FEM based on spectral elements (SE) is developed to improve the accuracy of the simulation of dynamic fractures (Liu et al., 2011).

5.2.1 Spectral Element-Based X-FEM

Spectral elements, originally proposed by Patera (1984) for fluid dynamics, have become popular for wave propagation (Capdeville et al., 2003; Seriani and Oliveira, 2007) because they are able to deliver significantly better accuracy as compared to standard finite element methods. This high-order numerical technique combines the high accuracy of spectral methods with the flexibility of finite elements. Furthermore, p-refinement of spectral elements is very easy to implement.

The spectral elements (SE) of Patera (1984) with a formulation are used here based on Karniadakis and Sherwin (1999). The element employs spaced nodes corresponding to the zeros of the Chebyshev or Legendre polynomials, called in this work a Chebyshev—Gauss point repartition. Taking the one-dimensional (1D) element as an example, the shape function can be expressed using Chebyshev polynomials as

$$N_I^{1D}(x) = \frac{(-1)^{I+p}(x^2 - 1)T_p'(x)}{c_I p^2 (x - x_I)} = \frac{2}{p}\sum_{J=0}^{p}\frac{1}{c_I c_J}T_J(x_I)T_J(x) \qquad (5.2)$$

where $c_I = 2$ if I is 0 or p; $c_I = 1$ otherwise. $T_p(x)$ is the pth order Chebyshev polynomial:

$$T_p(x) = \cos(p\theta) \quad \text{with} \quad \theta = \arccos x \quad \text{and} \quad x \in [-1, 1] \qquad (5.3)$$

Nodal coordinates in the element are chosen such that the derivative T_p' of T_p at x_I is zero:

$$T'_p(x_I) = 0, \quad I = 1, 2, ..., p - 1 \quad \text{and} \quad x_0 = -1, \quad x_p = 1 \qquad (5.4)$$

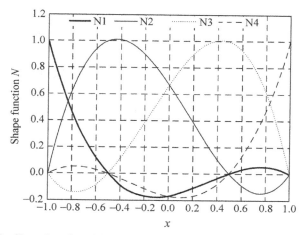

FIGURE 5.13 Shape function of 1D third-order spectral element.

One can then show that

$$x_I = -\cos\frac{I\pi}{p}, \quad \text{for } I \in \{0, 1, 2, \ldots, p\} \tag{5.5}$$

The shape functions of the four element nodes with $p = 3$ are shown in Figure 5.13.

In two dimensions (2D), the shape functions are obtained by a tensor product of the one-dimensional shape functions $N_I^{1D}(x)$:

$$N_{IJ}(\mathbf{x}) = N_{IJ}(x, y) = N_I^{1D}(x)N_J^{1D}(y) \tag{5.6}$$

Figure 5.14 shows the second- and third-order spectral elements and their nodes. It should be noted that the first-order spectral element ($p = 1$) degrades to a normal low-order element.

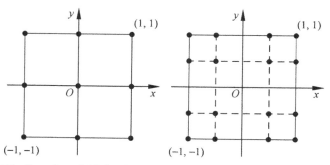

FIGURE 5.14 Second- and third-order spectral elements and node repartition.

According to the X-FEM formulation, Eq. (4.18) in Chapter 4, the displacement field in the spectral element-based X-FEM is expressed as

$$\mathbf{u}^{\mathrm{h}}(\mathbf{x}) = \sum_{I \in S} N_I^p(\mathbf{x})\mathbf{u}_I + \sum_{J \in S_{\mathrm{h}}} N_J^p(\mathbf{x})H(f(\mathbf{x}))\mathbf{a}_J(t) + \sum_{K \in S_{\mathrm{c}}} N_K^p(\mathbf{x})\Phi(\mathbf{x})\mathbf{b}_K(t) \quad (5.7)$$

where $N_J^p(\mathbf{x})$ is a p-order shape function, S is the total number of nodes in the model, S_{h} denotes the nodes of elements that are cut by the crack, and S_{c} is the set of nodes in the crack-tip influence domain. Compared to the low-order X-FEM, most of the formulations are the same except there is a difference in the shape function. So the spectral element-based X-FEM can be solved completely following the description in Chapter 4.

Since spectral elements have internal nodes (as shown in Figure 5.14), the stable time step in explicit analysis will decrease and needs to be studied

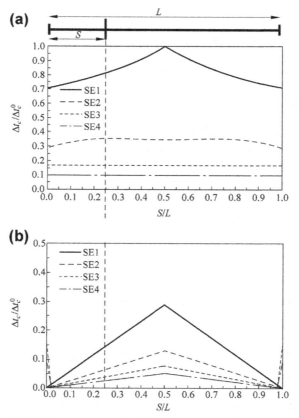

FIGURE 5.15 Critical time steps as a function of the location of the discontinuity in a 1D spectral element.

separately. We first consider a 1D cracked element of section A, length L, Young's modulus E, and mass density ρ as an example to decide the stable time step in this high-order X-FEM method. For a standard 1D two-node element (without discontinuity), the critical time step for the lumped mass matrix is L/c ($c = \sqrt{E/\rho}$ is wave velocity). This time step is chosen as the reference critical time step Δt_c^0 in the following. For both lumped mass and consistent mass, the critical time steps for the element at the crack location S of the discontinuity are given in Figure 5.15 for different spectral element orders. As shown in Figure 5.15, for higher order spectral elements with lumped mass, Δt_c does not change much with S, even when the discontinuity approaches either end of the element. But for the consistent mass, the critical time step Δt_c for higher order spectral elements decreases to zero as the discontinuity approaches one end of the element, so the stability of the explicit central difference scheme for the mesh cannot be guaranteed.

Next, we present several numerical examples to demonstrate the application of spectral element-based X-FEM in dynamic fracture simulations under the assumption of plane strain two-dimensional elasticity.

5.2.2 Mixed-Mode Static Crack

In this subsection, we consider the problem analyzed by Lee and Freund (1990) in which a semi-infinite plate with an edge notch is subjected to an impact velocity V_0. The problem is mixed mode due to the traction-free nature on the crack faces. The mode I stress intensity factor becomes negative when the lower part of the specimen comes into contact with the upper part.

The geometry of the finite plate is shown in Figure 5.16. The dimensions are as follows: $H = 6$ m, $L = 4$ m, crack length $a = 1$ m. A mesh of 60×119 spectral elements is used to define the model. The prescribed velocity on the boundary is $V_0 = 16.5$ m/s. Since the analytical solution is for an infinite plate,

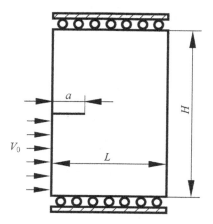

FIGURE 5.16 Semi-infinite plate with a stationary edge crack under mixed-mode loading.

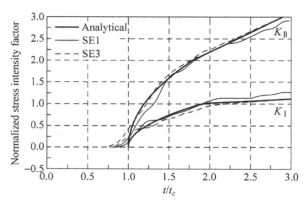

FIGURE 5.17 Normalized stress intensity factors as a function of time: analytical and numerical results.

it is only valid for a finite plate until the reflected wave arrives at the crack tip. Therefore, in our numerical solution we limit the simulation time to $t \leq 3t_c = 3a/c_1$, where c_1 is dilatational wave speed. The material properties of the linear elastic media are: Young's modulus $E = 200$ GPa, Poisson's ratio $\nu = 0.25$, and density $\rho = 7833$ kg/m^3. The stress intensity factors are normalized by the factor $\left(- EV_0 \sqrt{a/\pi} \right) / \left[2c_1 \left(1 - \nu^2 \right) \right]$. The analytical solutions for both stress intensity factors as a function of time are available for this problem (Lee and Freund, 1990).

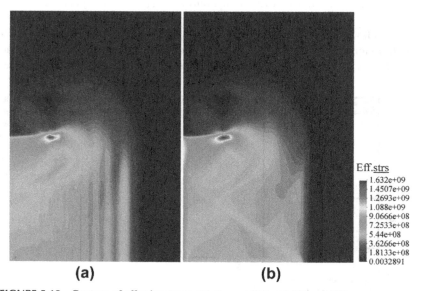

FIGURE 5.18 Contour of effective stress. (a) Element SE1. (b) Element SE3.

Figure 5.17 presents the normalized stress intensity factors K_{I} and K_{II} as a function of time for elements SE1 and SE3. We find that the element SE3 performs well during the whole loading history. In contrast, the results for element SE1 strongly deviate from the analytical results, especially at the late stages of the simulation. It can be also found from the contour of effective stress in Figure 5.18 that there are evident numerical perturbations near the crack tip in element SE1. The high-order spectral element is more accurate than the low-order element for this mixed-mode crack problem.

5.2.3 Kalthoff's Experiment

This example deals with a crack initially in mode II, based on experiments by Kalthoff (1985) and Böhme and Kalthoff (1982). A plate with two symmetrical edge cracks is impacted by a projectile at speed V_0. The two cracks are centered with respect to the specimen's geometry and their separation corresponds to the diameter of the projectile. The crack propagates with an overall angle from 60° to 70°. We chose $V_0 = 16.5$ m/s as a typical velocity for brittle fracture. A schematic description of the problem and the geometry are given in Figure 5.19. The dimensions of the specimen are: $L = 100$ mm, $l = 50$ mm, $a = 50$ mm, and the thickness is 16.5 mm. The material properties of the linear elastic media are: Young's modulus $E = 190$ GPa, Poisson's ratio $\nu = 0.3$, density $\rho = 8000$ kg/m^3, and fracture toughness $K_{\mathrm{IC}} = 68$ MPa$\cdot$m$^{1/2}$. This problem is widely used to validate numerical methods in dynamic crack propagation.

Regular meshes (38×38, 78×78) are employed in the simulations. The crack velocity is given by

$$
\dot{a} =
\begin{cases}
0 & \text{if } K_{\theta\theta} < K_{\mathrm{IC}}, \\
c_{\mathrm{r}}\left(1 - \left(\dfrac{K_{\mathrm{IC}}}{K_{\theta\theta}}\right)^2\right) & \text{otherwise}
\end{cases}
\tag{5.8}
$$

where c_{r} is Rayleigh wave speed and $K_{\theta\theta}$ is the equivalent dynamic stress intensity factor (see Chapter 2 for details).

Figure 5.20 shows the results for the fine mesh 78×78. It can be seen that the crack paths for elements SE2 and SE3 are almost identical; thus, that path can be considered as a converged solution. The angle is close to 70°.

The results for the coarse mesh 38×38 are given in Figure 5.21(a). The crack path of high-order element SE3 is still close to the angle of 70°, but not as smooth as that of fine mesh. The crack paths for different element meshes and element orders are compared in Figure 5.21(b). We can see that the crack path of element SE3 in the coarse mesh is even more accurate than that of element SE1 in the fine mesh.

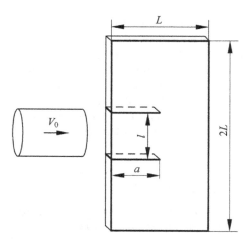

FIGURE 5.19   Geometry of Kalthoff's experiment.

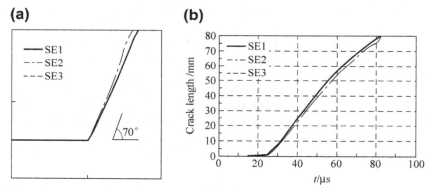

FIGURE 5.20   Crack paths (a) and crack lengths (b) of the Kalthoff test for element mesh 78 × 78.

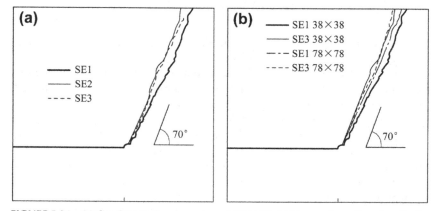

FIGURE 5.21   (a) Crack paths for element mesh 38 × 38. (b) Comparison of crack paths for different element meshes and element orders.

5.2.4 Mode I Moving Crack

In this subsection, the example in section 5.1.1 will be recalculated using high-order X-FEM. Figure 5.22 presents the normalized static stress intensity factors as a function of time. The stress intensity factors are normalized by $\sigma_0 \sqrt{h}$ and compared for elements SE1, SE2, and SE3. One can notice that apparent numerical oscillations appear in the stress intensity factor for element SE1. However, for elements SE2 and SE3, the stress intensity factors vary smoothly when the tensile wave and the reflected wave approach the crack tip, and the results are close to the analytical solutions.

Next we study the effect of a moving crack on the accuracy of the computed stress intensity factor and verify the performance of spectral elements in the simulation of dynamic crack propagation. A smoothed crack velocity is imposed after $1.5t_c$ as given below:

$$\dot{a} = \begin{cases} 0 & t < 1.5t_c \\ V_0 \sin\left[\dfrac{(t - 1.5t_c)\pi}{1.4t_c}\right] & 1.5t_c \leq t \leq 2.2t_c \\ V_0 & t > 2.2t_c \end{cases} \qquad (5.9)$$

Figure 5.23 presents the normalized stress intensity factors as a function of time. The stress intensity factors are normalized by $\sigma_0 \sqrt{h}$ and compared for elements SE1, SE2, and SE3. The numerical oscillations during crack propagation are strongly amplified in element SE1, as shown in Figure 5.23. However, the stress wave due to releasing the tip element is effectively decreased in elements SE2 and SE3. Figure 5.24 presents the relative errors in the stress intensity factor corresponding to the results shown in Figure 5.23.

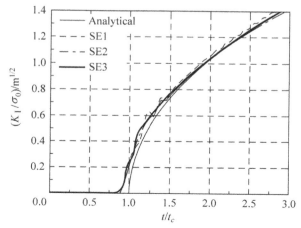

FIGURE 5.22   Normalized mode I stress intensity factor for a static crack as a function of time: analytical and numerical results.

Extended Finite Element Method

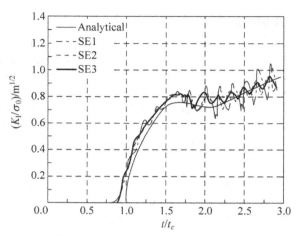

FIGURE 5.23 Normalized mode I stress intensity factor for a propagating crack as a function of time: analytical and numerical results.

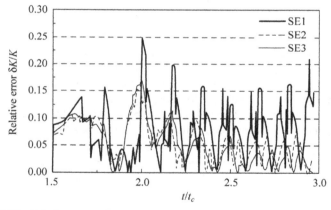

FIGURE 5.24 Relative errors in stress intensity factor during $1.5t_c < t < 3t_c$ for three spectral elements.

It underlines that spectral elements markedly improve the results. The small oscillations in elements SE2 and SE3 correspond to the crack tip passing from one element to the next.

The stress contours at $t = 3t_c$ for elements SE1 and SE3 are given in Figure 5.25. The crack-tip stress field distribution in element SE1 is strongly disordered by wave oscillations during the crack propagation, as shown in Figure 5.25. That is why large oscillations appear in the stress intensity factor of element SE1, as seen in Figure 5.23. This example clearly demonstrates the improvement in the simulation of dynamic crack propagation using spectral elements.

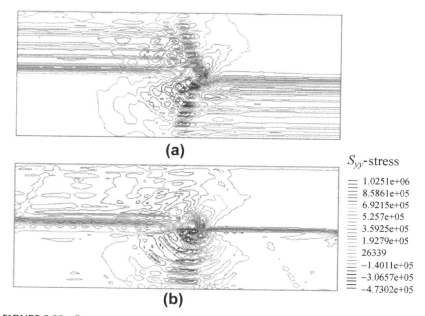

(a)

S_{yy}-stress

$\equiv$ 1.0251e+06
$\equiv$ 8.5861e+05
6.9215e+05
5.257e+05
3.5925e+05
1.9279e+05
26339
$\equiv$ −1.4011e+05
$\equiv$ −3.0657e+05
$\equiv$ −4.7302e+05

(b)

FIGURE 5.25   Stress contour at $t = 3t_c$. (a) Element SE1. (b) Element SE3.

5.3 CRACK BRANCHING SIMULATION

Crack branching is a phenomenon frequently observed in conjunction with dynamic fracture in brittle materials. How to numerically model the dynamic branching still remains a challenge in computational fracture mechanics. Daux et al. (2000) developed a methodology that constructs the enriched approximation based on the interaction of the discontinuous geometric features and calculated the stress intensity factor of static cracks with multiple branches. This can be used for static branched cracks, but the propagation of the crack and the dynamic branching process are not taken into consideration. Belytschko et al. (2003a) introduced loss of hyperbolicity as the criterion for dynamic crack propagation, and X-FEM has been applied to several dynamic crack growth problems including crack branching. In this simulation, the element in which the branching occurs is deleted simply to deal with the branching process. Song et al. (2006) simulated crack branching using the phantom nodes method, which is an alternative formulation of standard X-FEM, in which the stress in elements where three cracks have formed is set to zero for simplicity. In the author's group, a modified enrichment function is constructed using X-FEM to study the problem of dynamic crack branching, which is presented in the following subsections.

5.3.1 Crack Branching Enrichment

In the previous sections, the enrichment scheme of a single crack in a two-dimensional body was introduced; when considering a branched crack, the enrichment scheme should be improved. Daux et al. (2000) introduced discontinuous functions for modeling branched cracks. A junction function was used for the enrichment. Based on this idea, a shifted form was adopted by the authors (Xu et al., 2013) and the advantages are shown in the following.

When crack branching occurs, there are different situations that need to be considered. Here, we consider two kinds of new discontinuous elements, namely element with two cracks and elements with junctions, as shown in Figure 5.26. In these new kinds of enriched elements, the approximate displacement field needs to be modified, and a new description for the additional crack is required.

If two cracks exist in the same element, obviously one set of enriched degrees of freedom is not enough to describe the discontinuous feature of the element. Another set of enrichment functions needs to be introduced to express the additional discontinuity in the element. Meanwhile, another signed displacement function is also needed to describe the position of the second crack. The displacement field is expressed as

$$
\begin{aligned}
\mathbf{u}^{h}(\mathbf{X}) = & \sum_{I=1}^{4} N_I(\mathbf{X})\mathbf{u}_I + \sum_{J=1}^{4} N_J(\mathbf{X})\big(H\big(f^{\mathrm{I}}(\mathbf{X}) - f^{\mathrm{I}}(\mathbf{X}_J)\big)\big)\mathbf{q}_J^{\mathrm{I}} \\
& + \sum_{K=1}^{4} N_K(\mathbf{X})\big(H\big(f^{\mathrm{II}}(\mathbf{X}) - f^{\mathrm{II}}(\mathbf{X}_K)\big)\big)\mathbf{q}_K^{\mathrm{II}}
\end{aligned}
\tag{5.10}
$$

where $f^{\mathrm{I}}(\mathbf{X})$ and $f^{\mathrm{II}}(\mathbf{X})$ are the signed distance functions to the first crack (upper one in Figure 5.26, left) and second crack (lower one in Figure 5.26, left) respectively. $\mathbf{q}_J^{\mathrm{I}}$ and $\mathbf{q}_K^{\mathrm{II}}$ are the enriched degrees of freedom. In this formula we join the enrichments for the two disjointed cracks together, calculate the two distance functions and then the enriched functions respectively. By doing this the two strong discontinuities in the same element can be described together.

Another situation is where the element contains a junction where crack branching occurs. The number of degrees of freedom is the same as for the element with two cracks, and the enrichment scheme is similar. Besides the

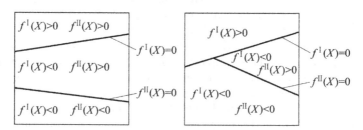

FIGURE 5.26 Distance funtions. Left: element with two cracks; right: element with junction.

original enrichment, a junction function is defined in this element. The displacement field is

$$
\mathbf{u}^h(\mathbf{X}) = \sum_{I=1}^{4} N_I(\mathbf{X})\mathbf{u}_I + \sum_{J=1}^{4} N_J(\mathbf{X})\big(H\big(f^{\mathrm{I}}(\mathbf{X}) - f^{\mathrm{I}}(\mathbf{X}_J)\big)\big)\mathbf{q}_J^{\mathrm{I}}
$$
$$
+ \sum_{K=1}^{4} N_K(\mathbf{X})\big(J(\mathbf{X}) - J(\mathbf{X}_K)\big)\mathbf{q}_K^{\mathrm{II}}
$$

(5.11)

where $J(\mathbf{X})$ is the junction function, defined as

$$
J(\mathbf{X}) = \begin{cases} H(f^{\mathrm{I}}(\mathbf{X})), & \text{on the side without branching} \\ H(f^{\mathrm{II}}(\mathbf{X})), & \text{on the side with branching} \end{cases}
$$

(5.12)

We regard the main crack and one of the two branches as the first crack (upper one in Figure 5.26, right), and $f^{\mathrm{I}}(\mathbf{X})$ is the signed distance function corresponding to it, while we take the other branch as the second crack (lower one in Figure 5.26, right), and $f^{\mathrm{II}}(\mathbf{X})$ is the signed distance function corresponding to it. It is worth mentioning that in the definition of $J(\mathbf{X})$, the side with branching refers to the region where $f^{\mathrm{I}}(\mathbf{X}) < 0$ shown in Figure 5.26. By applying the new enrichment to the X-FEM procedure given in Chapter 4, cracks with branches can be modeled.

5.3.2 Branch Criteria

The theoretical solution of the displacement and stress fields near the tip of a fast moving crack was given by Yoffe (1951). In theory, the stress field distribution has a relationship with the crack speed V. For a moving mode I crack, the main terms of the stress field are

$$
\sigma_{xx} = \frac{K_I(V)}{\sqrt{2\pi}} \frac{1 + \alpha_2^2}{4\alpha_1\alpha_2 - \left(1 + \alpha_2^2\right)^2} \left[\left(1 + 2\alpha_1^2 - \alpha_2^2\right)\frac{\cos\frac{\theta_1}{2}}{\sqrt{r_1}} - \frac{4\alpha_1\alpha_2}{1 + \alpha_2^2}\frac{\cos\frac{\theta_2}{2}}{\sqrt{r_2}} \right]
$$

(5.13)

$$
\sigma_{yy} = \frac{K_I(V)}{\sqrt{2\pi}} \frac{1 + \alpha_2^2}{4\alpha_1\alpha_2 - \left(1 + \alpha_2^2\right)^2} \left[-\left(1 + \alpha_2^2\right)\frac{\cos\frac{\theta_1}{2}}{\sqrt{r_1}} + \frac{4\alpha_1\alpha_2}{1 + \alpha_2^2}\frac{\cos\frac{\theta_2}{2}}{\sqrt{r_2}} \right]
$$

(5.14)

$$
\sigma_{xy} = \frac{K_I(V)}{\sqrt{2\pi}} \frac{1 + \alpha_2^2}{4\alpha_1\alpha_2 - \left(1 + \alpha_2^2\right)^2} \left[2\alpha_1\left(\frac{\sin\frac{\theta_1}{2}}{\sqrt{r_1}} - \frac{\sin\frac{\theta_2}{2}}{\sqrt{r_2}}\right) \right]
$$

(5.15)

where $\alpha_1 = (1 - V^2/c_1^2)^{1/2}$, $\alpha_2 = (1 - V^2/c_2^2)^{1/2}$ (c_1 and c_2 are the dilatational and shear wave speed respectively), $\tan\theta_1 = \alpha_1\tan\theta$, $\tan\theta_2 = \alpha_2\tan\theta$, $r_1 = r\cos\theta/\cos\theta_1$, and $r_2 = r\cos\theta/\cos\theta_2$. From the expression, the relationship between the stress and crack speed is clearly revealed. In Chapter 3 we

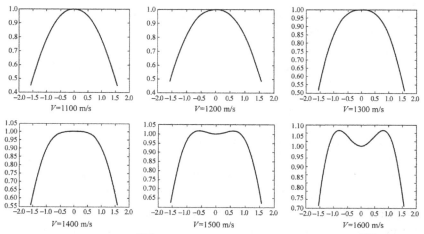

FIGURE 5.27   Curves of $\sigma_\theta \sqrt{2\pi r}/K_{\mathrm{I}} \sim \theta$ at varying propagation speeds (y: $\sigma_\theta \sqrt{2\pi r}/K_{\mathrm{I}}$; x: θ).

emphasize that the circumferential stress plays an important role in the crack propagation process. By coordinate transfer, we can obtain the circumferential stress:

$$
\sigma_\theta = \frac{K_I(V)}{\sqrt{2\pi}} \frac{1+\alpha_2^2}{4\alpha_1\alpha_2 - \left(1+\alpha_2^2\right)^2} \sin^2\theta \left[\left(1+2\alpha_1^2 - \alpha_2^2\right) \frac{\cos\frac{\theta_1}{2}}{\sqrt{r_1}} - \frac{4\alpha_1\alpha_2}{1+\alpha_2^2} \frac{\cos\frac{\theta_2}{2}}{\sqrt{r_2}} \right]
$$
$$
+ \frac{K_I(V)}{\sqrt{2\pi}} \frac{1+\alpha_2^2}{4\alpha_1\alpha_2 - \left(1+\alpha_2^2\right)^2} \cos^2\theta \left[-\left(1+\alpha_2^2\right) \frac{\cos\frac{\theta_1}{2}}{\sqrt{r_1}} + \frac{4\alpha_1\alpha_2}{1+\alpha_2^2} \frac{\cos\frac{\theta_2}{2}}{\sqrt{r_2}} \right]
$$
$$
+ \frac{K_I(V)}{\sqrt{2\pi}} \frac{1+\alpha_2^2}{4\alpha_1\alpha_2 - \left(1+\alpha_2^2\right)^2} (-2\sin\theta\cos\theta)
$$

$$(5.16)$$

In order to qualitatively analyze this relationship, the curves of $\sigma_\theta \sqrt{2\pi r}/K_{\mathrm{I}} \sim \theta$ for different propagation speeds are plotted in Figure 5.27 for a given material.

For a mode I crack, if the crack propagates quasi-statically, the maximum circumferential stress will stay in the original direction of the initial crack which is $\theta = 0$ in this case. For a dynamically propagating crack, the situation is different. Figure 5.27 apparently reveals a change of the normalized circumferential stress distribution for different crack speeds, and we can qualitatively find that when the velocity is low, the maximum normalized circumferential stress occurs in the direction of $\theta = 0$, which means the crack will grow along the original direction, while as the velocity increases, the maximum normalized circumferential stress occurs along two symmetrical directions. Based on the viewpoint of the maximum circumferential stress propagation criterion, we suppose that during the acceleration process of a mode I dynamic crack, the departure of the maximum circumferential stress from the direction of $\theta = 0$ explains the crack branching.

The theory agrees with the fact that dynamic crack branching is related to the crack speed. When the propagation speed is given, it provides a criterion for whether the crack will branch or not. From the analysis of the normalized circumferential stress, we can find the direction in which the maximum stress occurs. If the speed is not high enough, the direction will be $\theta = 0$, otherwise we can get a nonzero angle, which is regarded as the branching angle. Though this is not a physical criterion, it can provide an available criterion in the numerical calculation to provide an approximate simulation.

5.3.3 Numerical Examples

In this subsection, both static branched cracks and dynamic crack branching are simulated to check the capability and accuracy of the method described above.

(a) Static Example

The first example is a static edge crack with branches in a finite plate. As shown in Figure 5.28, the right edge of the plate is fixed, and uniform tensions are applied at the top and bottom edges. The properties of the

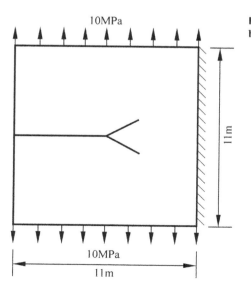

FIGURE 5.28 An edge crack with branches.

10MPa

11m

10MPa

11m

FIGURE 5.29 Mises stress distribution (units: MPa).

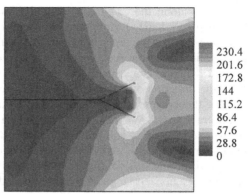

```
230.4
201.6
172.8
144
115.2
86.4
57.6
28.8
0
```

material are $E = 210$ GPa and $\nu = 0.3$. With a structure mesh and an X-FEM algorithm, the deformed configuration and Mises stress field are obtained, as shown in Figure 5.29. From the distribution of Mises stress, we can find that there are stress concentrations on both of the two crack tips, while at the junction of the three cracks the stress level is low, because it is on the free surface.

Then by considering a static branched crack in an infinite plate under uniaxial traction, as shown in Figure 5.30, a quantitative study is carried out in which the stress intensity factors of the three crack tips are calculated and compared with analytical results. The geometric parameters in the simulation are $\theta = 45°$ and the length ratio $b/a = 1.0$.

In the stress intensity factor handbook, we can find the theoretical SIFs for this problem. The SIFs of the cracks are calculated by the following formulae:

$$K_{\mathrm{IA}} = F_{\mathrm{IA}}\sigma\sqrt{\pi\frac{c}{2}}, \quad K_{\mathrm{IIA}} = F_{\mathrm{IIA}}\sigma\sqrt{\pi\frac{c}{2}} \tag{5.17}$$

FIGURE 5.30 Geometry of a branched crack in an infinite plate.

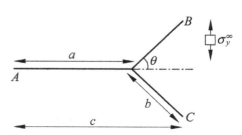

TABLE 5.1 Coefficients of Stress Intensity Factors

	Theoretical	Numerical	Error (%)
F_{IA}	1.044	1.033	1.05
F_{IIA}	0	4.916×10^{-4}	—
F_{IB}	0.5	0.488	2.49
F_{IIB}	−0.5	−0.483	3.40
F_{IC}	0.5	0.489	2.20
F_{IIC}	0.5	0.483	3.40

$$K_{IB} = F_{IB}\sigma\sqrt{\pi\frac{c}{2}}, \quad K_{IIB} = F_{IIB}\sigma\sqrt{\pi\frac{c}{2}} \tag{5.18}$$

$$K_{IC} = F_{IC}\sigma\sqrt{\pi\frac{c}{2}}, \quad K_{IIC} = F_{IIC}\sigma\sqrt{\pi\frac{c}{2}} \tag{5.19}$$

where $c = a + b\cos\theta$.

In the calculations, we use the method of interaction integral to calculate the stress intensity factors. Details on the interaction integral can be found in Chapter 3. Coefficient F is calculated from the SIF, and the numerical results are compared with the theoretical values, as shown in Table 5.1, where we can see that the numerical results agree well with the theoretical solutions.

(b) Dynamic Branching

Another example concerns the propagation of a crack in a pre-notched specimen, as shown in Figure 5.31. The dimensions of the specimen are 0.1 m × 0.04 m, and the length of the initial crack is 0.05 m. The properties of the material are $E = 32$ GPa, $\nu = 0.2$, and $\rho = 2450$ kg/m³. The dilatational wave speed is $c_d = 3809.5$ m/s, the shear wave speed $c_s = 2332.8$ m/s, and the Rayleigh wave

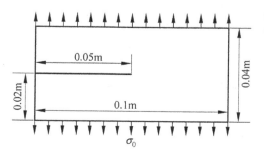

FIGURE 5.31 Schematic of the crack branching model.

speed $c_R = 2119.0$ m/s. Tensile tractions $\sigma = 1.0$ MPa are applied at both the top and bottom edges as a step function in time. In this simulation, quadrilateral elements are used. Numerical results for this problem have been given by Belytschko et al. (2003a), Song et al. (2006), and Borden et al. (2012).

In this simulation, the maximum circumferential stress criterion is used to obtain the crack growth paths and the theory introduced in subsection 5.3.2 is used to predict crack branching. The solution given by Yoffe is established under some assumptions that are very difficult to satisfy in practice. Here we use it as an estimation of the initiation of branching.

The propagation paths and the Mises stress fields obtained from the simulation are shown in Figure 5.32. The crack starts to propagate at about 8 μs, and it is a typical mode I crack. The initial crack grows along the original direction until the branching criterion is reached. When crack branching takes

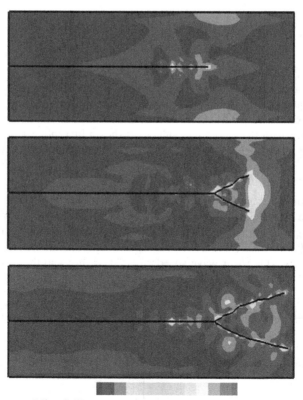

Mises/MPa　1　2　3　4　5　6　7　8　9　10　11　12

FIGURE 5.32 **Crack branching paths (from top to bottom corresponding to times 25, 35, and 45 μs).**

place, the single crack becomes two branches, which occurs at about 25 μs. After that the two branches go on to propagate separately.

5.4 SUMMARY

In this chapter, 2D X-FEM is applied to simulate static cracks, mixed-mode crack propagation, interaction between crack and voids, dynamic crack branching, etc., as well as other two-dimensional discontinuity problems. The precision of the simulation results is discussed, and then the reliability and robustness are verified for the X-FEM program developed by the authors and colleagues.

The results of mixed-mode crack propagation are compared with experimental data. The propagation angles agree with experimental data and prove the capability and efficiency of the algorithm and program in capturing complex cracks. By constructing the absolute value shape function, the material interface is simulated independently of the mesh, and the results compare well with commercial software analysis. In addition, the interaction of void and crack is studied under the X-FEM framework, to reveal the influence of voids on the stress intensity factor and the mixed degree of fracture modes. Two-dimensional high-order X-FEM is developed based on the spectral element method, which can be used to control the numerical disturbances that occur when simulating dynamic fracture by low-order elements. Finally, two types of new enriched elements, namely elements containing two cracks and elements with junctions, are constructed to effectively model the crack branching process.

Chapter 6

X-FEM on Continuum-Based Shell Elements

Chapter Outline

6.1 Introduction 104
6.2 Overview of Plate and Shell Fracture Mechanics 104
  6.2.1 Kirchhoff Plate and Shell Bending Fracture Theory 105
  6.2.2 Reissner Plate and Shell Bending Fracture Theory 109
6.3 Plate and Shell Theory Applied In Finite Element Analysis 113
6.4 Brief Introduction to General Shell Elements 115
  6.4.1 Belytschko—Lin—Tsay Shell Element 115
  6.4.2 Continuum-Based Shell Element 116
6.5 X-FEM on CB Shell Elements 119
  6.5.1 Shape Function of a Crack Perpendicular to the Mid-Surface 119
  6.5.2 Shape Function of a Crack Not Perpendicular to the Mid-Surface 123

  6.5.3 Total Lagrangian Formulation 125
  6.5.4 Time Integration Scheme and Linearization 127
  6.5.5 Continuum Element Transformed to Shell 128
6.6 Crack Propagation Criterion 129
  6.6.1 Stress Intensity Factor Computation 129
  6.6.2 Maximum Energy Release Rate Criterion 133
6.7 Numerical Examples 136
  6.7.1 Mode I Central Through-Crack in a Finite Plate 136
  6.7.2 Mode III Crack Growth in a Plate 137
  6.7.3 Steady Crack in a Bending Pipe 137
  6.7.4 Crack Propagation Along a Given Path in a Pipe 139
  6.7.5 Arbitrary Crack Growth in a Pipe 140
6.8 Summary 140

Extended Finite Element Method. http://dx.doi.org/10.1016/B978-0-12-407717-1.00006-6

6.1 INTRODUCTION

In engineering applications, various kinds of shell structures are widely used, such as sheet metal in an automobile, the fuselage, wings and rudder of aircraft, walls of pressure vessels, architectural domes, electronic device shells, etc. The characteristic of a shell structure is that the dimension of thickness is far less than the other two dimensions of length and width. The configuration of the shell's middle layer is usually a curved surface. If the mid-surface is flat, the shell degenerates to a plate structure. Because of the complexity of force and deformation in the plate and shell structures, their theoretical solutions are limited in an infinite dimension or geometric symmetry, as well as under simple loads (e.g., Timoshenko and Woinowsky-Krieger, 1959; Hwang et al., 1988). For geometric and load complexity problems, the numerical simulation by the FEM is the most efficient method of analysis (e.g., Cantin and Clough, 1968; Ashwell and Sabir, 1972; Landau et al., 1978; Bathe and Ho, 1981).

Compared with the theory of plates and shells, fracture mechanics theory seems to be very limited. It can be seen in section 6.2 that two most accepted theories so far have essentially been developed for plate problems, but are only suitable for the special case of infinite dimensions and symmetry boundary conditions (e.g., Liu and Jiang, 2000). Therefore, the computational method of fracture mechanics for plates and shells is leading the way in computational mechanics (e.g., Zhuang, 1995; Mohan, 1998; Colombo and Giglio, 2006; Kim et al., 2010). In section 6.3, the general adopted computational formula is discussed. The continuum-based (CB) shell element for X-FEM is established in section 6.4. It includes the use of CB shells to deal with complex cracks, which may be not perpendicular to the mid-surface. The calculation method using stress intensity factors is also set up in section 6.5 for X-FEM with CB shells. This method is important in solving complex fracture problems in middle thickness shells.

6.2 OVERVIEW OF PLATE AND SHELL FRACTURE MECHANICS

Plate and shell supported transverse loads are widely used structures in engineering. The fracture mechanics of plates and shells is mainly delineated into two theories (Liu and Jiang, 2000): Kirchhoff–Love plate and shell (hereafter referred to as Kirchhoff plate and shell) bending fracture theory and Mindlin–Reissner plate and shell (hereafter referred to as Reissner plate and shell) bending fracture theory. The former is suitable for thin plates and shells, while the latter is suitable for thicker plates and shells. Since an approximate boundary condition of equivalent shear force is introduced into Kirchhoff plates and shells, there is a serious theoretical drawback when it is used to deal with bending fracture problems. Compared with Reissner plate and shell, the

Reissner theory has much higher precision, but the solution is very complex. We will introduce the essential concept and analytical methodology for these two fracture shell theories, as well as the differences and the relation between them. It is supposed that the reader has a fundamental knowledge of plate and shell theory.

6.2.1 Kirchhoff Plate and Shell Bending Fracture Theory

The Kirchhoff theory is suitable for thin plates and shells in a small deflection situation. Thin plates and shells are those with a ratio of thickness h and characteristic dimension b of the mid-surface of about $h/b < 1/5-1/8$. Small deflection indicates that the deflection is far less than the thickness of the plate, normally limited by $w/h < 1/5$.

There are three basic assumptions for plate bending deformation in Kirchhoff theory:

1. The straight line segment, which is perpendicular to the mid-surface before deformation, remains linear after deformation and is still perpendicular to the deformed mid-surface. The influence of elongation along the mid-surface to deflection may be neglected.
2. The points located in the mid-surface have no in-plane displacement.
3. There is no inter-compression between the layers parallel to the mid-surface.

As shown in Figure 6.1, a line segment AB, which is perpendicular to the mid-surface before deformation, moves to the position A′B′ after deformation. The displacement component at an arbitrary point on the segment AB is given by

$$\begin{cases} u(x,y,z) = -z\beta = -z\dfrac{\partial \omega}{\partial x} \\[2mm] v(x,y,z) = -z\dfrac{\partial \omega}{\partial y} \\[2mm] \omega(x,y,z) = \omega(x,y) \end{cases} \tag{6.1}$$

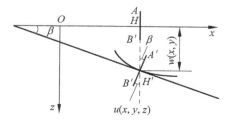

FIGURE 6.1   Kirchhoff plate and shell deformation (Liu and Jiang, 2000).

According to the geometric equation and constitutive relation, the internal force of the plate and shell repressed by deflection can be obtained as

$$
\begin{cases}
M_x = \int\limits_{-h/2}^{h/2} \sigma_x z \, dz = -D\left(\dfrac{\partial^2 \omega}{\partial x^2} + \nu \dfrac{\partial^2 \omega}{\partial y^2}\right) \\[3mm]
M_y = \int\limits_{-h/2}^{h/2} \sigma_y z \, dz = -D\left(\dfrac{\partial^2 \omega}{\partial y^2} + \nu \dfrac{\partial^2 \omega}{\partial x^2}\right) \\[3mm]
M_{xy} = \int\limits_{-h/2}^{h/2} \tau_{xy} z \, dz = -D(1-\nu)\dfrac{\partial^2 \omega}{\partial x \partial y} \\[3mm]
Q_x = \int\limits_{-h/2}^{h/2} \tau_{xz} z \, dz = -D\dfrac{\partial}{\partial x}\left(\dfrac{\partial^2 \omega}{\partial x^2} + \dfrac{\partial^2 \omega}{\partial y^2}\right) \\[3mm]
Q_y = \int\limits_{-h/2}^{h/2} \tau_{yz} z \, dz = -D\dfrac{\partial}{\partial y}\left(\dfrac{\partial^2 \omega}{\partial x^2} + \dfrac{\partial^2 \omega}{\partial y^2}\right)
\end{cases}
\tag{6.2}
$$

where $D = Eh^3/[12(1-\nu^2)]$ is bending stiffness of the plate and h represents thickness. E and ν represent elastic modulus and Poisson's ratio respectively. The balance equations of a plate expressed by the internal forces are

$$
\begin{cases}
\dfrac{\partial M_x}{\partial x} + \dfrac{\partial M_{xy}}{\partial y} - Q_x = 0 \\[3mm]
\dfrac{\partial M_{xy}}{\partial x} + \dfrac{\partial M_y}{\partial y} - Q_y = 0 \\[3mm]
\dfrac{\partial Q_x}{\partial x} + \dfrac{\partial Q_y}{\partial y} + q = 0
\end{cases}
\tag{6.3}
$$

The terms of shear force are eliminated from the internal balance equations (6.3), and the terms in equations (6.2) are substituted into equations (6.3). Finally, the basic equations of Kirchhoff theory expressed by deflection can be obtained as follows:

$$
\frac{\partial^4 \omega}{\partial x^4} + 2\frac{\partial^4 \omega}{\partial x^2 \partial y^2} + \frac{\partial^4 \omega}{\partial y^4} = \frac{q}{D}
\tag{6.4}
$$

There are three kinds of boundary conditions:

1. Fixed boundary: $w = \overline{w}$, $\partial w / \partial n = \overline{\alpha}$.
2. Simple supported boundary: $w = \overline{w}$, $M_n = \overline{M_n}$.
3. Free boundary: $M_n = \overline{M_n}$, $M_{ns} = \overline{M_{ns}}$, $Q_n = \overline{Q_n}$.

To solve Eq. (6.4), only two boundary conditions are required. Since there are three conditions in free boundary (3), obviously one more condition is necessary. Therefore, the equivalent shear $V_n = \partial M_{ns}/\partial s + Q_n$ is introduced into Kirchhoff theory. The second and third conditions in (3) are equivalent to one relation, which results in the free boundary (3) only consisting of two boundary conditions, such that:

4. Free boundary: $M_n = \overline{M}_n$, $V_n = \overline{V}_n$.

In fact, it is from this equivalence to the free boundary that an essential drawback in Kirchhoff plate theory emerges as regards solving fracture problems. The equivalence at the free boundary condition reduces the precision of the solution of the stress field near the free boundary. The crack surface is constituted by two free boundaries, thus it affects the solution process and the precision of the stress field at the crack tip.

In order to study the bending crack tip field of a Kirchhoff thin plate, the rectangular Cartesian coordinate system and polar coordinate system are introduced as shown in Figure 6.2, where the crack tip is located at the original position. The transformation relation of two coordinates is

$$x = r \cos \theta, \quad y = r \sin \theta \tag{6.5}$$

The bending differential equation of Kirchhoff plate theory is adopted in the same form as the governing equation of the stress function in plane problems. Therefore, a set of successful methods used in plane fracture problems, such as the complex variable functions method, integration transformation method, locality-integration method, etc., can all be used in the analysis of Kirchhoff plate bending fracture problems. The closed-form solutions have been obtained for a number of typical problems, as shown in Figure 6.3. For an infinite plate including a crack with length $2a$ under uniform loading far away from the crack, the stress intensity factor of bending fracture is given by:

1. Under uniform bending moment far away from crack M^{∞}:

$$K_{\mathrm{I}} = \frac{6\sqrt{\pi a}}{h^2} M^{\infty} \tag{6.6}$$

2. Under uniform torsion moment far away from crack M_{xy}^{∞}:

$$K_{\mathrm{I}} = 0, \quad K_{\mathrm{II}} = \frac{6\sqrt{\pi a}}{h^2} M_{xy}^{\infty} \tag{6.7}$$

where h is the thickness of the plate.

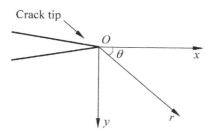

FIGURE 6.2   Rectangular Cartesian coordinate system and polar coordinate system at the crack tip position.

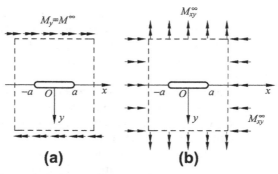

FIGURE 6.3 Two typical plate bending fracture problems.

Kirchhoff thin plate theory has been widely used in engineering, and Kirchhoff bending fracture theory has also subsequently been developed. The stress intensity factor formulae for Kirchhoff plate bending cracks are included in some stress intensity factor manuals. There are similar fracture analysis methods between shell and plate, but the analysis of shells is rather more complex than that of plates. Systematic research on this subject can be found in Liu and Jiang (2000).

For mode I and mode II mixed cracks, the stress field at the crack tip of a plate can be found using the series development method:

$$
\sigma_x = \frac{3+\nu}{2(1-\nu)} \cdot \frac{z}{h} \cdot \frac{K_1^{(K)}}{\sqrt{2\pi r}} \left(-3\cos\frac{\theta}{2} - \cos\frac{5}{2}\theta \right)
$$

$$
+ \frac{3+\nu}{2(1-\nu)} \cdot \frac{z}{h} \cdot \frac{K_2^{(K)}}{\sqrt{2\pi r}} \left(-\frac{9+7\nu}{1-\nu}\sin\frac{\theta}{2} + \sin\frac{5}{2}\theta \right) + O(r^0)
$$

$$
\sigma_y = \frac{3+\nu}{2(1-\nu)} \cdot \frac{z}{h} \cdot \frac{K_1^{(K)}}{\sqrt{2\pi r}} \left(\frac{11+5\nu}{1-\nu}\cos\frac{\theta}{2} + \cos\frac{5}{2}\theta \right)
$$

$$
+ \frac{3+\nu}{2(1-\nu)} \cdot \frac{z}{h} \cdot \frac{K_2^{(K)}}{\sqrt{2\pi r}} \left(\sin\frac{\theta}{2} - \sin\frac{5}{2}\theta \right) + O(r^0)
$$

$$
\tau_{xy} = \frac{3+\nu}{2(1-\nu)} \cdot \frac{z}{h} \cdot \frac{K_1^{(K)}}{\sqrt{2\pi r}} \left(-\frac{7+\nu}{1-\nu}\sin\frac{\theta}{2} - \sin\frac{5}{2}\theta \right)
$$

$$
+ \frac{3+\nu}{2(1-\nu)} \cdot \frac{z}{h} \cdot \frac{K_2^{(K)}}{\sqrt{2\pi r}} \left(\frac{5-3\nu}{1-\nu}\cos\frac{\theta}{2} - \cos\frac{5}{2}\theta \right) + O(r^0)
$$

$$
\tau_{xz} = \frac{h^2 - 4z^2}{4h(3+\nu)\sqrt{2\pi r}r} \left(-K_1^{(K)}\cos\frac{3\theta}{2} + K_2^{(K)}\sin\frac{3\theta}{2} \right)
$$

$$
\tau_{yz} = \frac{h^2 - 4z^2}{4h(3+\nu)\sqrt{2\pi r}r} \left(K_1^{(K)}\sin\frac{3\theta}{2} + K_2^{(K)}\cos\frac{3\theta}{2} \right)
$$

(6.8)

From this solution, some obvious drawbacks of Kirchhoff theory can be observed:

1. Although the singularity order about r for in-plane stress is $r^{-1/2}$, which is the same as the plane problem, the angle distribution law is different. It cannot degenerate to the plane problem.
2. The singularity order about r for transverse shear stress is $r^{-3/2}$, which is not consistent with the physical reality.

6.2.2 Reissner Plate and Shell Bending Fracture Theory

There are three basic assumptions in Reissner thicker plate theory:

1. The straight line segment, which is perpendicular to the mid-surface before deformation, remains linear straight after deformation and does not need to be perpendicular to the deformed mid-surface. The elongation along the length direction may be neglected.
2. The points located in the mid-surface have no in-plane displacement.
3. There is no inter-compression between the layers parallel to the mid-surface.

Compared with Kirchhoff theory, we can see that the maximum improvement of Reissner plate theory is that it relaxes the deformation restraint. It allows the line segment, which is perpendicular to the mid-surface before deformation, to be rotated after deformation, as shown in Figure 6.4. A line segment AB, which is perpendicular to the mid-surface before deformation, allows rotation relative to the mid-surface and moves to the position A''B'' after deformation. It is supposed that the line A''B'' rotates an angle φ_x relative to the line AB in the Oxz plane, and the same line in the Oyz plane rotates another angle φ_y relative to the line AB. Then, the displacement component of an arbitrary point in the plate can be

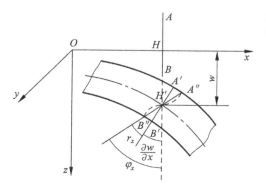

FIGURE 6.4 Reissner shell deformation (Liu and Jiang, 2000).

indicated by a deflection ω and the other two rotation angles φ_x and φ_y respectively, such that:

$$\begin{cases} u(x,y,z) = -z\varphi_x(x,y) \\ v(x,y,z) = -z\varphi_y(x,y) \\ \omega(x,y,z) = \omega(x,y) \end{cases} \tag{6.9}$$

According to the geometrical equation and constitutive relation, the internal forces of the plate can be expressed by ω, φ_x and φ_y as follows:

$$\begin{cases} M_x = \int_{-h/2}^{h/2} \sigma_x z \mathrm{d}z = -D\left(\frac{\partial \varphi_x}{\partial x} + \nu \frac{\partial \varphi_y}{\partial y}\right) \\[2mm] M_y = \int_{-h/2}^{h/2} \sigma_y z \mathrm{d}z = -D\left(\frac{\partial \varphi_y}{\partial y} + \nu \frac{\partial \varphi_x}{\partial x}\right) \\[2mm] M_{xy} = \int_{-h/2}^{h/2} \tau_{xy} z \mathrm{d}z = -\frac{1}{2}D(1-\nu)\left(\frac{\partial \varphi_x}{\partial y} + \frac{\partial \varphi_y}{\partial x}\right) \\[2mm] Q_x = \int_{-h/2}^{h/2} \tau_{xz} z \mathrm{d}z = C\left(\frac{\partial \omega}{\partial x} - \varphi_x\right) \\[2mm] Q_y = \int_{-h/2}^{h/2} \tau_{yz} z \mathrm{d}z = C\left(\frac{\partial \omega}{\partial y} - \varphi_y\right) \end{cases} \tag{6.10}$$

where $D = Eh^3/[12(1-\nu^2)]$ is the bending stiffness of the plate, $C = \frac{5}{6}\mu h$ is transverse shear stiffness, and μ is shear modulus. The balance equations of the plate expressed using internal forces are given by

$$\begin{cases} \dfrac{\partial M_x}{\partial x} + \dfrac{\partial M_{xy}}{\partial y} - Q_x = 0 \\[3mm] \dfrac{\partial M_{xy}}{\partial x} + \dfrac{\partial M_y}{\partial y} - Q_y = 0 \\[3mm] \dfrac{\partial Q_x}{\partial x} + \dfrac{\partial Q_y}{\partial y} + q = 0 \end{cases} \tag{6.11}$$

Substituting Eq. (6.10) into Eq. (6.11), the balance equations of a plate can be expressed by deflection ω, and the other rotation angles φ_x and φ_y, such that

$$
\begin{cases}
D\left(\dfrac{\partial^2 \varphi_x}{\partial x^2} + \dfrac{1-\nu}{2}\dfrac{\partial^2 \varphi_x}{\partial y^2} + \dfrac{1+\nu}{2}\dfrac{\partial^2 \varphi_y}{\partial x \partial y}\right) + C\left(\dfrac{\partial \omega}{\partial x} - \varphi_x\right) = 0 \\[2mm]
D\left(\dfrac{1+\nu}{2}\dfrac{\partial^2 \varphi_x}{\partial x \partial y} + \dfrac{1-\nu}{2}\dfrac{\partial^2 \varphi_y}{\partial x^2} + \dfrac{\partial^2 \varphi_y}{\partial y^2}\right) + C\left(\dfrac{\partial \omega}{\partial y} - \varphi_y\right) = 0 \qquad (6.12) \\[2mm]
C\left(\dfrac{\partial^2 \omega}{\partial x^2} + \dfrac{\partial^2 \omega}{\partial y^2} - \dfrac{\partial \varphi_x}{\partial x} - \dfrac{\partial \varphi_y}{\partial y}\right) + q = 0
\end{cases}
$$

The typical boundary conditions of Reissner plate theory are provided by

1. Fixed boundary: $w = \overline{w}$, $\varphi_n = \overline{\varphi_n}$, $\varphi_s = \overline{\varphi_s}$.
2. The first kind of simple supported boundary: $w = \overline{w}$, $\varphi_s = \overline{\varphi_s}$, $M_n = \overline{M_n}$.
3. The second kind of simple supported boundary: $w = \overline{w}$, $M_n = \overline{M_n}$, $M_{ns} = \overline{M_{ns}}$.
4. Free boundary: $M_n = \overline{M_n}$, $M_{ns} = \overline{M_{ns}}$, $Q_n = \overline{Q_n}$.

For the Reissner thicker plate with a crack, as shown in Figure 6.2, the rectangular Cartesian coordinate system and polar coordinate system are set up, where the crack tip is located at the original position. All of the stress components near the bending crack tip have the singularity order about $r^{-1/2}$. However, the angle distribution function is not related to the external load and configuration of the plate. Therefore, in accordance with the definitions in fracture mechanics for three kinds of stress intensity factors, such as opening mode, slipping mode, and tearing mode, the stress intensity factor at the bending crack tip of a Reissner plate can be calculated, such that

$$
\left.
\begin{aligned}
K_{\mathrm{I}}(z) &= \lim_{r \to 0} \sqrt{2\pi r}\,\sigma_\theta(r, 0, z) \\
K_{\mathrm{II}}(z) &= \lim_{r \to 0} \sqrt{2\pi r}\,\tau_{r\theta}(r, 0, z) \\
K_{\mathrm{III}}(z) &= \lim_{r \to 0} \sqrt{2\pi r}\,\tau_{\theta z}(r, 0, z)
\end{aligned}
\right\} \qquad (6.13)
$$

where z is a coordinate along the normal line starting from the mid-surface of the plate. The stress components are given by

$$
\sigma_\theta = \frac{12 M_\theta z}{h^3}, \qquad \tau_{r\theta} = \frac{12 M_{r\theta} z}{h^3}, \qquad \tau_{\theta z} = \frac{Q_\theta}{h^3}\left(h^2 - 4z^2\right) \qquad (6.14)
$$

The difference of stress intensity factor at the crack tip location compared with the in-plane and antiplane problem is that it varies along the thickness in a Reissner plate. The distribution law depends on the relevant stress situation. The maximum stress intensity factors of opening mode I, K_{I}, and slipping mode II, K_{II}, occur on the surfaces of the plate, such that $z = \pm h/2$, while the tearing mode III, K_{III}, occurs in the mid-surface, such that $z = 0$.

It can be found that the definitions of three kinds of stress intensity factors at the crack tip location for a Reissner plate are the same as the corresponding definitions of the stress intensity factors for plane and antiplane cracks. There are the same orders of stress singularity near the crack tip as shown in Eq. (6.13), and the angle distribution functions are the same too. The differences of stress intensity factors between them are that there are uniform distributions along the thickness orientation in the plane and antiplane crack tip fields. The bending crack tip field of a Reissner plate is quite complex, such that the stress fields of mode I and mode II cracks are distributed linearly along the thickness and have maximum values located at the top and bottom surface layers, while the stress field of a mode III crack presents a parabolic distribution along the thickness and has its maximum value in the mid-surface of the plate. Their stress distributions are similar to the distribution laws corresponding to the bending normal stress and shear stress of the plate. These characteristics indicate that it may be convenient to study the coupling problem of plate bending and in-plane stresses. The more precise 3D analysis demonstrates the scientific feature of fracture analysis using Reissner plate theory.

For a mode I, mode II, and mode III mixed crack, the stress singularity field at the crack tip in Reissner plate theory can be obtained by the series development method:

$$
\begin{aligned}
\sigma_x = {} & \frac{K_1^{(R)}}{\sqrt{2\pi r}} \cdot \frac{2z}{h} \cos\frac{\theta}{2}\left(1 - \sin\frac{\theta}{2}\sin\frac{3}{2}\theta\right) \\
& - \frac{K_2^{(R)}}{\sqrt{2\pi r}} \cdot \frac{2z}{h} \sin\frac{\theta}{2}\left(2 + \cos\frac{\theta}{2}\cos\frac{3}{2}\theta\right) + O\!\left(r^0\right) \\[2mm]
\sigma_y = {} & \frac{K_1^{(R)}}{\sqrt{2\pi r}} \cdot \frac{2z}{h} \cos\frac{\theta}{2}\left(1 + \sin\frac{\theta}{2}\sin\frac{3}{2}\theta\right) \\
& + \frac{K_2^{(R)}}{\sqrt{2\pi r}} \cdot \frac{2z}{h} \sin\frac{\theta}{2}\cos\frac{\theta}{2}\cos\frac{3}{2}\theta + O\!\left(r^0\right) \\[2mm]
\tau_{xy} = {} & \frac{K_1^{(R)}}{\sqrt{2\pi r}} \cdot \frac{2z}{h} \cos\frac{\theta}{2}\sin\frac{\theta}{2}\cos\frac{3}{2}\theta \\
& - \frac{K_2^{(R)}}{\sqrt{2\pi r}} \cdot \frac{2z}{h} \cos\frac{\theta}{2}\left(1 - \sin\frac{\theta}{2}\sin\frac{3}{2}\theta\right) + O\!\left(r^0\right) \\[2mm]
\tau_{xz} = {} & -\frac{K_3^{(R)}}{\sqrt{2\pi r}} \cdot \frac{h^2 - 4z^2}{h^2} \sin\frac{\theta}{2} + O\!\left(r^0\right) \\[2mm]
\tau_{yz} = {} & \frac{K_3^{(R)}}{\sqrt{2\pi r}} \cdot \frac{h^2 - 4z^2}{h^2} \cos\frac{\theta}{2} + O\!\left(r^0\right)
\end{aligned}
\tag{6.15}
$$

It can be seen from Eq. (6.15), no matter comparison with the in-plane and antiplane problems, that the solutions of Reissner plate theory have the same $r^{-1/2}$ order singularity and angle distribution function. During the fracture analysis of plates and shells, therefore, Reissner plate theory is more precise than Kirchhoff plate theory.

6.3 PLATE AND SHELL THEORY APPLIED IN FINITE ELEMENT ANALYSIS

In terms of stress singularity solutions at the crack tip location, Reissner plate theory is more reasonable than Kirchhoff plate theory. However, in the larger domain including the crack tip, the two theories each have their individual dominant areas. A series of works by Hui (1993) and Viz (1995) showed that Reissner plate theory is dominant in the domain very close to the crack tip ($r < h$); in the domain far away from the crack tip ($h < r \ll a$), Kirchhoff plate theory dominates, as shown in Figure 6.5.

Quantitative research by Zucchini et al. (2000) has studied this phenomenon in detail. In their work, 3D FEM is used to compute the stress field at the crack tip for an infinite plate including a penetrating crack. The simulation results have been compared with the solutions given by Kirchhoff and Reissner theories. Three loading cases were studied separately: the plate under bending, shearing, and twisting. The FEM results were compared with the solutions of two theories for the plate under bending, as shown in Figure 6.6. In addition, the FEM results were compared with the solutions of Kirchhoff theory for the plate under shear loading, as shown in Figure 6.7, and the FEM results were compared with the solutions of Reissner theory for the plate under twist loading, as shown in Figure 6.8.

A comparison of these results shows indeed that the solutions of Reissner theory agree well with the FEM results for a small region of crack tip location

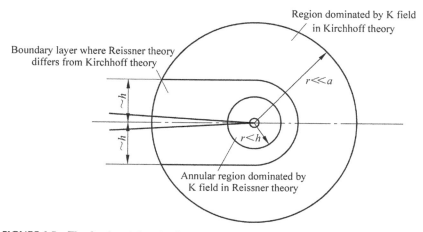

FIGURE 6.5 The dominant domains for two plate and shell theories (Zucchini et al., 2000).

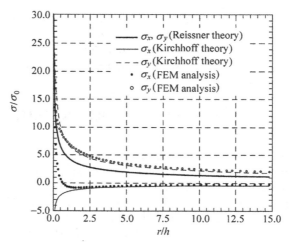

FIGURE 6.6 FEM results compared with solutions of two theories for a plate under bending (Zucchini et al., 2000).

within $r < 0.1h$, while for a large area outside of this small region, the solutions of Kirchhoff theory agree well with the FEM results. The main reason is the ratio between the thickness of the plate and the computational radius of the crack tip location. The former tends to the thicker plate and the later tends to the thin plate.

We learn from this work that it is not precise to construct the shape function of a shell element for the crack tip singularity field of X-FEM by applying the Kirchhoff or Reissner theories. Since the dimension of the mid-surface is much larger than the thickness of the shell element, both Kirchhoff and Reissner theories have their respective shortcomings, and could not be

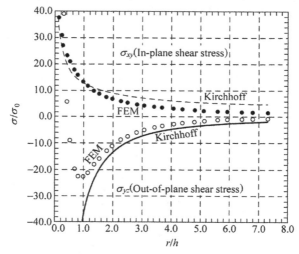

FIGURE 6.7 FEM results compared with solutions of Kirchhoff theory for a plate under shear loading (Zucchini et al., 2000).

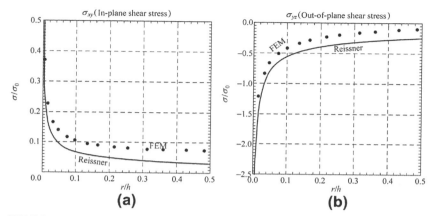

FIGURE 6.8   FEM results compared with solutions of Reissner theory for a plate under twist loading (Zucchini et al., 2000).

used if the shape function is constructed to satisfy the crack tip singularity field in such a large region. The problem is how to use two theoretical models to develop an integration X-FEM program. Beyond that, both theories have a more essential limitation, such that they are developed only for the plane plate; they are not capable of dealing with the shell element problem with curvature. Therefore, it is necessary to find a more suitable solution method.

6.4 BRIEF INTRODUCTION TO GENERAL SHELL ELEMENTS

The X-FEM formatting of two kinds of shell elements is discussed: Belytschko—Lin—Tsay and continuum-based (CB) shell elements. The former shell element was developed by Belytschko et al. in 2004, when they first established the shell element formula of X-FEM. The latter X-FEM CB shell element was developed by Zhuang and Cheng in 2011 based on the CB shell format. Also, the computational formula of the stress intensity factor is introduced for 3D cracks. In the following subsections, the computational formulae of the two kinds of shell element is described.

6.4.1 Belytschko—Lin—Tsay Shell Element

The Belytschko—Lin—Tsay shell element is a commonly used element in engineering computations, and has high efficiency in simulations. However, because one-point integration is adopted, the hourglass mode may occur in the element, which is an obvious shortcoming. Hourglass control is applied in practice. Based upon this element, the shell element formula of X-FEM was first established by Belytschko.

The Belytschko—Lin—Tsay shell element is a four-node quadrilateral element. The mid-surface of the element consists of four corner nodes, as shown in Figure 6.9. In order to get a convenient calculation of the

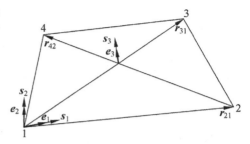

FIGURE 6.9 Local coordinate system of Belytschko–Lin–Tsay shell element.

stress–strain relationship, a set of a local coordinate system is defined in the interior of element$< \tilde{e}_1, \tilde{e}_2, \tilde{e}_3 >$, where

$$\tilde{e}_3 = \frac{s_3}{\|s_3\|}$$

$$s_3 = r_{31} \times r_{42}$$

$$s_1 = r_{21} - (r_{21} \cdot \tilde{e}_3)\tilde{e}_3 \qquad (6.16)$$

$$\tilde{e}_1 = \frac{s_1}{\|s_1\|}$$

$$\tilde{e}_2 = \tilde{e}_3 \times \tilde{e}_1$$

During the deformation process, $\tilde{e}_3$ with fiber orientation f must satisfy the following condition:

$$\left| \hat{e}_3 \cdot f - 1 \right| < \delta \qquad (6.17)$$

In engineering computations, δ is generally of the order of 10^{-2}. This limitation of the element is that it can only be undertaken for small deformation.

The velocity at every point in the shell element is given by

$$v = v^m - \tilde{z}e_3 \times \omega \qquad (6.18)$$

where v^m is the velocity at the mid-surface of shell, $\tilde{z}$ is thickness of fiber direction, and ω is angle velocity relative to the mid-surface at the point.

6.4.2 Continuum-Based Shell Element

The theory of continuum-based (CB) shell elements was first proposed by Ahmad et al. (1970), and it was extended to nonlinear problems by Hughes and Liu (1981a, b). Later, Simo and Fox (1989), Buechter and Ramm (1992), and Parisch (1995) developed this theory. The CB shell element establishes the kinematical formulation of the shell element on a 3D solid element (Belytschko et al., 2000). It has some basic assumptions:

1. The fiber remains linear (modified Mindlin–Reissner assumption).
2. Stress along the shell thickness direction is zero (plane stress assumption).

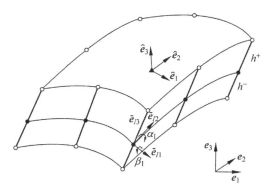

FIGURE 6.10 The CB shell element in the 3D case with eight nodes and 5 DOFs at each node.

3. The momentum results from fiber elongation, and the momentum balance is ignored along the fiber direction.

Next, the CB shell element developed by Zhuang and Cheng (2011c) is described for eight three-dimensional master nodes with 5 DOFs at each node, as illustrated in Figure 6.10.

The configuration of an eight-node CB shell element is illustrated in Figure 6.10. $< e_1, e_2, e_3 >$ is a global coordinate system. The black nodes are master nodes that pertain to the shell element and the white nodes are slave nodes that pertain to the solid element. The master and slave nodes are connected by fibers (the thick lines). By establishing the local nodal coordinates at the master node, the unit base vectors $\bar{e}_1$, $\bar{e}_2$, and $\bar{e}_3$ are perpendicular to each other and $\bar{e}_3$ coincides with the fiber orientation. The DOFs of each slave node pair can be translated into three translation DOFs of the master node and two rotation DOFs, α and β, of the fiber about the axes $\bar{e}_1$ and $\bar{e}_2$ respectively. If the upper and lower thicknesses of the fiber are h^+ and h^- respectively, then the translation matrix can be written as

$$
T = \begin{bmatrix}
1 & 0 & 0 & h^+ \cos(\bar{e}_{I1}, e_1) & -h^+ \cos(\bar{e}_{I2}, e_1) \\
0 & 1 & 0 & h^+ \cos(\bar{e}_{I1}, e_2) & -h^+ \cos(\bar{e}_{I2}, e_2) \\
0 & 0 & 1 & h^+ \cos(\bar{e}_{I1}, e_3) & -h^+ \cos(\bar{e}_{I2}, e_3) \\
1 & 0 & 0 & -h^- \cos(\bar{e}_{I1}, e_1) & h^- \cos(\bar{e}_{I2}, e_1) \\
0 & 1 & 0 & -h^- \cos(\bar{e}_{I1}, e_2) & h^- \cos(\bar{e}_{I2}, e_2) \\
0 & 0 & 1 & -h^- \cos(\bar{e}_{I1}, e_3) & h^- \cos(\bar{e}_{I2}, e_3)
\end{bmatrix}
\tag{6.19}
$$

We can correlate the degrees of freedom of the master and slave nodes:

$$
\begin{Bmatrix}
u_i^+ \\
v_i^+ \\
w_i^+ \\
u_i^- \\
v_i^- \\
w_i^-
\end{Bmatrix}
= T
\begin{Bmatrix}
u_i \\
v_i \\
w_i \\
\alpha_i \\
\beta_i
\end{Bmatrix}
\tag{6.20}
$$

In order to introduce the second assumption, for a plane stress condition, laminar coordinates $< \widehat{e}_1, \widehat{e}_2, \widehat{e}_3 >$ should be established at each integral point. In laminar coordinates, the 3D elastic constitutive law is

$$\left\{ \begin{array}{c} \widehat{\sigma}_{xx} \\ \widehat{\sigma}_{yy} \\ \widehat{\tau}_{xy} \\ \widehat{\tau}_{xz} \\ \widehat{\tau}_{yz} \\ \widehat{\sigma}_{zz} \end{array} \right\} = C \left\{ \begin{array}{c} \widehat{\varepsilon}_{xx} \\ \widehat{\varepsilon}_{yy} \\ \widehat{\gamma}_{xy} \\ \widehat{\gamma}_{xz} \\ \widehat{\gamma}_{yz} \\ \widehat{\varepsilon}_{zz} \end{array} \right\} = \left[\begin{array}{cc} \overline{C}_{5\times5} & \overline{C}_{5\times1} \\ \overline{C}_{1\times5} & \overline{C}_{1\times1} \end{array} \right] \left\{ \begin{array}{c} \widehat{\varepsilon}_{xx} \\ \widehat{\varepsilon}_{yy} \\ \widehat{\gamma}_{xy} \\ \widehat{\gamma}_{xz} \\ \widehat{\gamma}_{yz} \\ \widehat{\varepsilon}_{zz} \end{array} \right\} \qquad (6.21)$$

in which the constitutive matrix is

$$C = \frac{E(1-\nu)}{(1+\nu)(1-2\nu)} \begin{bmatrix} 1 & \dfrac{\nu}{1-\nu} & \dfrac{\nu}{1-\nu} & 0 & 0 & 0 \\ \dfrac{\nu}{1-\nu} & 1 & \dfrac{\nu}{1-\nu} & 0 & 0 & 0 \\ \dfrac{\nu}{1-\nu} & \dfrac{\nu}{1-\nu} & 1 & 0 & 0 & 0 \\ 0 & 0 & 0 & \dfrac{1-2\nu}{2(1-\nu)} & 0 & 0 \\ 0 & 0 & 0 & 0 & \dfrac{1-2\nu}{2(1-\nu)} & 0 \\ 0 & 0 & 0 & 0 & 0 & \dfrac{1-2\nu}{2(1-\nu)} \end{bmatrix}$$

$$(6.22)$$

Substituting the plane stress condition ($\widehat{\sigma}_{zz} = 0$) into Eq. (6.21), we can get the relation among the constituents of the elastic matrix C:

$$C'_{5\times5} = \overline{C}_{5\times5} - \overline{C}_{5\times1} \overline{C}_{1\times1} \overline{C}^T_{1\times5} \qquad (6.23)$$

Then the constitutive law becomes

$$\left\{ \begin{array}{c} \widehat{\sigma}_{xx} \\ \widehat{\sigma}_{yy} \\ \widehat{\sigma}_{xy} \\ \widehat{\sigma}_{xz} \\ \widehat{\sigma}_{yz} \end{array} \right\} = C'_{5\times5} \left\{ \begin{array}{c} \widehat{\varepsilon}_{xx} \\ \widehat{\varepsilon}_{yy} \\ \widehat{\gamma}_{xy} \\ \widehat{\gamma}_{xz} \\ \widehat{\gamma}_{yz} \end{array} \right\} \qquad (6.24)$$

Using coordinate transformation, the stress tensor obtained in laminar coordinates can be translated into global coordinates:

$$\sigma = R\hat{\sigma}R^T \qquad (6.25)$$

In each iterative step of the computation, the thicknesses of the fibers are updated as

$$
\begin{aligned}
h^+ &= \int_0^1 h_I^+ F_{\zeta\zeta}(+\zeta)\mathrm{d}\zeta \\
h^- &= \int_0^1 h_I^- F_{\zeta\zeta}(-\zeta)\mathrm{d}\zeta
\end{aligned}
\qquad (6.26)
$$

in which $F_{\zeta\zeta}(+\zeta)$ and $F_{\zeta\zeta}(-\zeta)$ are the deformation gradients above and below the mid-surface respectively. $F_{\zeta\zeta}$ can be found via $F_{\zeta\zeta} = F_{ij}(e_i \cdot \bar{e}_3)(e_j \cdot \bar{e}_3)$.

6.5 X-FEM ON CB SHELL ELEMENTS

The format of X-FEM is introduced into CB shell elements in this section. The CB shell element is convenient when considering the fiber elongation, including the thickness variation, and has very fine computation precision for thicker shells. It is also convenient to apply in thin shell problems (Belytschko et al., 2000). In addition, the deformation of CB shell elements is set up based on 3D solid elements, which can easily simulate the fractured model of the curved-surface shell, and the complicated situation of a crack that is not perpendicular to the mid-surface. The key point of this work is to construct the enriched shape function for 3D solid elements in order to capture the displacement fields precisely for the element crossed by a crack and element embedded by a crack tip. Then, the corresponding enriched DOFs are transformed into the master nodes of the shell elements. Another key point of this work is to introduce a new method for computing the stress intensity factor in the X-FEM of CB shell elements. According to this method, the stress intensity factors for mode I, mode II, and mode III can be directly calculated in a 3D domain, which is presented in section 6.6. Therefore, this method is more reasonable than the method of using the stress intensity factor of the plate to deal with the fractured model of the curved-surface shell.

6.5.1 Shape Function of a Crack Perpendicular to the Mid-Surface

In this work, each CB shell element has eight master nodes on the mid-surface based on standard shell elements and has 16 slave nodes based on solid elements, as shown in Figure 6.10. In order to satisfy the modified

Mindlin–Reissner assumption, only two slave nodes are assigned in each fiber direction connected to one master node. The shape functions of the 16-node solid element can be written by using the elemental coordinates (ξ, η, ζ):

$$N_1 = \frac{1}{8}(1+\zeta)\left[(1+\xi)(1+\eta) - (1-\xi^2)(1+\eta) - (1+\xi)(1-\eta^2)\right]$$

$$N_2 = \frac{1}{8}(1+\zeta)\left[(1-\xi)(1+\eta) - (1-\xi^2)(1+\eta) - (1-\xi)(1-\eta^2)\right]$$

$$N_3 = \frac{1}{8}(1+\zeta)\left[(1-\xi)(1-\eta) - (1-\xi^2)(1-\eta) - (1-\xi)(1-\eta^2)\right]$$

$$N_4 = \frac{1}{8}(1+\zeta)\left[(1+\xi)(1-\eta) - (1-\xi^2)(1-\eta) - (1+\xi)(1-\eta^2)\right]$$

$$N_5 = \frac{1}{4}(1+\zeta)(1-\xi^2)(1+\eta)$$

$$N_6 = \frac{1}{4}(1+\zeta)(1-\eta^2)(1-\xi)$$

$$N_7 = \frac{1}{4}(1+\zeta)(1-\xi^2)(1-\eta)$$

$$N_8 = \frac{1}{4}(1+\zeta)(1-\eta^2)(1+\xi)$$

$$N_9 = \frac{1}{8}(1-\zeta)\left[(1+\xi)(1+\eta) - (1-\xi^2)(1+\eta) - (1+\xi)(1-\eta^2)\right]$$

$$N_{10} = \frac{1}{8}(1-\zeta)\left[(1-\xi)(1+\eta) - (1-\xi^2)(1+\eta) - (1-\xi)(1-\eta^2)\right]$$

$$N_{11} = \frac{1}{8}(1-\zeta)\left[(1-\xi)(1-\eta) - (1-\xi^2)(1-\eta) - (1-\xi)(1-\eta^2)\right]$$

$$N_{12} = \frac{1}{8}(1-\zeta)\left[(1+\xi)(1-\eta) - (1-\xi^2)(1-\eta) - (1+\xi)(1-\eta^2)\right]$$

$$N_{13} = \frac{1}{4}(1-\zeta)(1-\xi^2)(1+\eta)$$

$$N_{14} = \frac{1}{4}(1-\zeta)(1-\eta^2)(1-\xi)$$

$$N_{15} = \frac{1}{4}(1-\zeta)(1-\xi^2)(1-\eta)$$

$$N_{16} = \frac{1}{4}(1-\zeta)(1-\eta^2)(1+\xi)$$

$$(6.27)$$

Only the situation of a crack perpendicular to the mid-surface is considered here. When the element is crossed by a crack, similar to the construction of the enriched shape function in 2D X-FEM as shown in Eq. (4.13), the enriched shape function in a 3D continuum element can be expressed by the following equation:

$$\phi_J(x) = N_J(x)s(x) \qquad (6.28)$$

where $s(x)$ is the signed distance function, which is defined as follows: it is positive when x lies on one side of the crack and negative when x lies on the other side.

In the writing program, the scheme below may be adopted: as shown in Figure 6.11, it is necessary to discuss the two cases for the cracked surface with kinking or no kinking in the element.

(1) The cracked surface has no kinking in the element: the line of intersection AB between the cracked surface and mid-surface of the mother element forms a vector along the crack orientation, which is in terms of t_c, as shown in Figure 6.11(a). The projection of point x to the mid-surface is a point x'. From the initial point of vector t_c, point A, to a point x' we have a vector r, then the signed function $s(x)$ may be defined as

$$s(x) = \text{sign}\left[r \cdot \left(\overrightarrow{O\zeta} \times t_c \right) \right] \qquad (6.29)$$

Dealing with the crack surface in this way, Figure 6.12 demonstrates two representative configurations of the enriched shape function for the eight-node element crossed by a crack (the thickness variation is ignored).

(2) The cracked surface has kinking in the element: the lines of intersection AB and BC between the cracked surface and mid-surface of the mother element form two vectors along the crack orientation, which are in terms

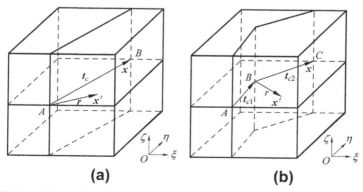

(a)   **(b)**

FIGURE 6.11   3D continuum element crossed by a crack perpendicular to the mid-surface. (a) Cracked surface with no kinking in the element. (b) Cracked surface with kinking in the element.

FIGURE 6.12 Two representative configurations of enriched shape function for an eight-node element crossed by a crack.

of t_{c1} and t_{c2} respectively, as shown in Figure 6.11(b). The projection of point x to the mid-surface is a point x'. From the point of intersection between vectors t_{c1} and t_{c2}, point B, to a point x' we have a vector r, then the signed function $s(x)$ may be defined by the following formula:

$$\left.\begin{aligned} s(x) &= 1, \quad r \cdot \left(\overrightarrow{O\zeta} \times t_{c1} \right) > 0 \quad \text{and} \quad r \cdot \left(\overrightarrow{O\zeta} \times t_{c2} \right) > 0 \\ s(x) &= -1, \quad \text{else} \end{aligned}\right\} \, t_{c2} \cdot \left(\overrightarrow{O\zeta} \times t_{c1} \right) > 0$$

$$\left.\begin{aligned} s(x) &= -1, \quad r \cdot \left(\overrightarrow{O\zeta} \times t_{c1} \right) < 0 \quad \text{and} \quad r \cdot \left(\overrightarrow{O\zeta} \times t_{c2} \right) < 0 \\ s(x) &= 1, \quad \text{else} \end{aligned}\right\} \, t_{c2} \cdot \left(\overrightarrow{O\zeta} \times t_{c1} \right) < 0$$

$$(6.30)$$

For the element embedded by a crack tip, it is necessary to set up the coordinates using the point of intersection B between the crack tip and mid-surface as the original point. The zero axial orientation of θ is coincident with the crack growth direction, as shown in Figure 6.13. The projection of point x to the mid-surface is a point x'. From the original point of polar coordinates, point B, to a point x' we have a vector r, then the enriched shape function $\phi_J(x)$ may be expressed in the following form:

$$\phi_J(x) = N_J(x)b(x) \tag{6.31}$$

FIGURE 6.13 3D continuum element embedded by a crack tip.

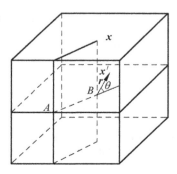

where $b(x)$ may be a set of linear combinations of the functions given below:

$$B(x) = \left[\sqrt{r}\sin\frac{\theta}{2}, \ \sqrt{r}\sin\frac{\theta}{2}\sin\theta, \ \sqrt{r}\cos\frac{\theta}{2}, \ \sqrt{r}\cos\frac{\theta}{2}\sin\theta \right] \qquad (6.32)$$

The set of functions given in Eq. (6.32) is in accordance with the enriched function of 2D X-FEM for the element embedded by a crack tip. In this way, the variation of displacement field variables at the crack tip is ignored along the thickness orientation. Belytschko et al. (2002) have used this shape function in solving 3D nonplane crack growth problems for continuum elements. The difference from their work is that the shape function is used for the continuum-based shell element here. The four representative configurations of enriched shape functions for the eight-node element embedded by a crack tip (the thickness variation is ignored) are shown in Figure 6.14.

6.5.2 Shape Function of a Crack Not Perpendicular to the Mid-Surface

The computational method for a crack perpendicular to the mid-surface of a shell element was described in subsection 6.5.1, which includes the construction of enriched shape functions for two cases, the element crossed by a crack and the element embedded by a crack tip. In this subsection, we discuss how to construct these two enriched shape functions for a crack not perpendicular to the mid-surface. The difference from the previous discussion lies in the requirement to estimate the relative position between the integral point and crack surface in the 3D spatial domain, while it is not sufficient only to use an integral point projected on to the mid-surface.

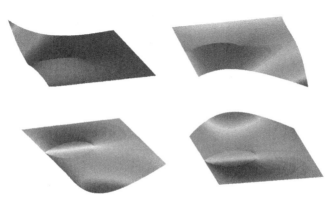

FIGURE 6.14   The four representative configurations of enriched shape functions for the eight-node element embedded by a crack tip.

The case of an element crossed by a crack is first discussed, as shown in Figure 6.15, which is divided into two situations, those with crack surface kinking and no kinking in the element.

(1) No crack surface kinking in the element. The line of intersection EF between the cracked surface ABCD and mid-surface of the mother element forms a vector along the crack orientation, which is given in terms of t_c, as shown in Figure 6.15(a). The line AD on the cracked surface and perpendicular to t_c forms another vector t_{oc}. The vector r is from point E to point x' and then the signed function $s(x)$ can be defined as

$$s(x) = \text{sign}[r \cdot (t_{oc} \times t_c)] \tag{6.33}$$

(2) Crack surface kinking in the element. The lines of intersection EI and IF between the cracked surface AGBCHD and mid-surface of the mother element form two vectors along the crack orientation, which are in terms of t_{c1} and t_{c2} respectively, as shown in Figure 6.15(b). The line AD on the cracked surface and perpendicular to t_c forms another vector t_{oc}. The vector r is from point E to point x' and then the signed function $s(x)$ can be defined as

$$\begin{aligned} s(x) = 1, \quad & r \cdot (t_{oc} \times t_{c1}) > 0 \quad \text{and} \quad r \cdot (t_{oc} \times t_{c2}) > 0 \\ s(x) = -1, \quad & \text{else} \end{aligned} \left. \right\} \quad t_{c2} \cdot (t_{oc} \times t_{c1}) > 0$$
$$\begin{aligned} s(x) = -1, \quad & r \cdot (t_{oc} \times t_{c1}) < 0 \quad \text{and} \quad r \cdot (t_{oc} \times t_{c2}) < 0 \\ s(x) = 1, \quad & \text{else} \end{aligned} \left. \right\} \quad t_{c2} \cdot (t_{oc} \times t_{c1}) < 0$$

$$\tag{6.34}$$

It can be found from the above equations that the crack perpendicular to the mid-surface is a special case of a crack that is not perpendicular to the

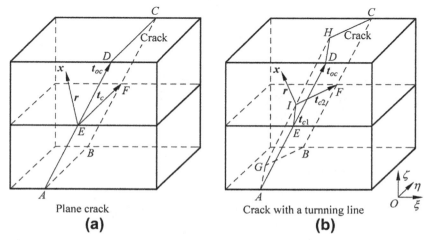

FIGURE 6.15 **3D continuum element with a crack not perpendicular to the mid-surface.** (a) Crack surface with no kinking in the element. (b) Crack surface kinking in the element.

mid-surface. Both Eqs. (6.33) and (6.34) are suitable for use in the situations mentioned in subsection 6.5.1. t_{oc} is parallel to the vector $\overrightarrow{O\zeta}$ when the crack is perpendicular to the mid-surface.

When the crack tip is embedded in the element, a cylinder coordinate is established along the crack tip, as shown in Figure 6.16. The $r-\theta$ plane of the cylinder coordinate is perpendicular to the plane of the crack tip and it does not necessarily coincide with the mid-surface. The enriched shape function is set up based on this cylinder coordinate. There is a coordinate transformation relation between the cylinder and the mother element. According to this relation, the derivative of the shape function with respect to the mother element coordinate can be obtained, given matrix B.

6.5.3 Total Lagrangian Formulation

In the previous deduction, the construction method of enriched shape functions has been established, which is a basic part of X-FEM with CB shell elements. The enriched shape functions are constructed on solid elements, so the formula of X-FEM is also established on general solid elements. After that, the discretized formulation of FEM is transformed onto the CB shell element by the transformation matrix. In this subsection, the total Lagrangian formulation of X-FEM is deduced.

The weak form of the total Lagragian formulation is similar to the standard FEM. For any $\delta u(X) \in u_0$, $u(X,t) \in u$, is the variational function space, in which $u_0 = \{\delta u(X) | \delta u(X) \in C^0, \ \delta u(X) = 0 \text{ on } \Gamma_0^u, \ \delta u \text{ is discontinuous on } \Gamma_0^c\}$; and $u = \{u(X,t) | u(X,t) \in C^0, \ u(X,t) = \overline{u} \text{ on } \Gamma_0^u, \ u \text{ is discontinuous on } \Gamma_0^c\}$ is the displacement trial function space. The following equations are found:

$$\int_{\Omega_0} (\delta F^T : P - \rho_0 \delta u \cdot b + \rho_0 \delta u \cdot \ddot{u}) d\Omega_0 - \sum_{i=1}^{n_{SD}} \int_{\Gamma_{t_i}^0} (\delta u \cdot e_i)\left(e_i \cdot \bar{t}_i^0\right) d\Gamma_0 = 0$$

$$\text{or} \quad \int_{\Omega_0} \left(\delta F_{ij} P_{ji} - \rho_0 \delta u_i b_i + \rho_0 \delta u_i \ddot{u}_i\right) d\Omega_0 - \sum_{i=1}^{n_{SD}} \int_{\Gamma_{t_i}^0} \delta u \bar{t}_i^0 d\Gamma_0 = 0$$

$$(6.35)$$

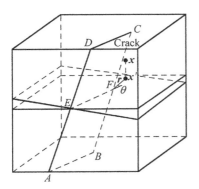

FIGURE 6.16   Element embedded by a crack tip.

It can also be written in terms of virtual work:

$$\delta w^{\text{int}}(\delta u, u) - \delta w^{\text{ext}}(\delta u, u) + \delta w^{\text{kin}}(\delta u, u) = 0 \tag{6.36}$$

where the internal virtual work is

$$\delta w^{\text{int}} = \int_{\Omega_0} \delta F^{\text{T}} : P d\Omega_0 = \int_{\Omega_0} \delta F_{ij} P_{ji} d\Omega_0 \tag{6.37}$$

the external virtual work is

$$\begin{aligned}
\delta w^{\text{ext}} &= \int_{\Omega_0} \rho_0 \delta u \cdot b d\Omega_0 + \sum_{i=1}^{n_{\text{SD}}} \int_{\Gamma_{t_i}^0} (\delta u \cdot e_i)\left(e_i \cdot \bar{t}_i^0\right) d\Gamma_0 \\
&= \int_{\Omega_0} \rho_0 \delta u_i b_i d\Omega_0 + \sum_{i=1}^{n_{\text{SD}}} \int_{\Gamma_{t_i}^0} \delta u_i \bar{t}_i^0 d\Gamma_0
\end{aligned} \tag{6.38}$$

and the inertial force virtual work is

$$\delta w^{\text{kin}} = \int_{\Omega_0} \rho_0 \delta u \cdot \ddot{u} d\Omega_0 = \int_{\Omega_0} \rho_0 \delta u_i \ddot{u}_i d\Omega_0 \tag{6.39}$$

By substituting the enriched shape functions constructed previously for the element crossed by a crack and the element embedded by a crack tip into the discretized FEM space, the displacement function is given by

$$\begin{aligned}
u(X,t) &= N_I(X)u_I(t) + \phi_J(X)u_J^*(t) \\
\text{or} \quad u_i(X,t) &= N_I(X)u_{iI}(t) + \phi_J(X)u_{iJ}^*(t)
\end{aligned} \tag{6.40}$$

and the variational function is

$$\begin{aligned}
\delta u(X,t) &= N_I(X)\delta u_I(t) + \phi_J(X)\delta u_J(t) \\
\text{or} \quad \delta u_i(X,t) &= N_I(X)\delta u_{iI}(t) + \phi_J(X)\delta u_{iJ}^*(t)
\end{aligned} \tag{6.41}$$

By introducing the discretized displacement function into Eqs. (6.37) and (6.38) respectively, the expressions for internal and external node forces can be obtained:

$$\begin{aligned}
f_{iI}^{\text{int}} &= \int_{\Omega_0} \left(B_{Ij}^0\right)^T P_{ji} d\Omega_0 = \int_{\Omega_0} \frac{\partial N_I}{\partial X_j} P_{ji} d\Omega_0 \\
q_{iI}^{\text{int}} &= \int_{\Omega_0} \left(B_{Ij}^{*0}\right)^T P_{ji} d\Omega_0 = \int_{\Omega_0} \frac{\partial \phi_I}{\partial X_j} P_{ji} d\Omega_0
\end{aligned} \tag{6.42}$$

$$\begin{aligned}
f_{iI}^{\text{ext}} &= \int_{\Omega_0} N_I \rho_0 b_i d\Omega_0 + \int_{\Gamma_{t_i}^0} N_I \bar{t}_i^0 d\Gamma_0 \\
q_{iI}^{\text{ext}} &= \int_{\Omega_0} \phi_I \rho_0 b_i d\Omega_0 + \int_{\Gamma_{t_i}^0} \phi_I \bar{t}_i^0 d\Gamma_0
\end{aligned} \tag{6.43}$$

Since the discontinuity displacement field in the element is constructed using the enriched shape function, the second equations in Eqs. (6.42) and (6.43) are the new internal force and external force respectively, contributing to the element original node.

We have a similar 2D situation as seen in section 4.6, and the diagonal mass matrix is used in the program. In the element that is crossed by a crack, the mass on either side of the crack is calculated separately and is applied to the relevant nodes. By combining the internal force, external force, and mass matrix, as well as considering the condition of zero virtual work as expressed in Eq. (6.36), the discretized momentum equations are finally obtained:

$$
\begin{aligned}
M_{ijIJ}\ddot{u}_{jJ} + f_{iI}^{\text{int}} &= f_{iI}^{\text{ext}} \\
M_{ijIJ}^{*}\ddot{a}_{jJ} + q_{iI}^{\text{int}} &= q_{iI}^{\text{ext}}
\end{aligned} \tag{6.44}
$$

The second equation in Eq. (6.44) is the balancing condition at element nodes, which results in a discontinuous displacement field in the element. It can be found that the discontinuity in the element is included since the new DOF is attached.

6.5.4 Time Integration Scheme and Linearization

Equation (6.44) is a semidiscretized equation because it is only discretized in the spatial domain. In the solution of time integration, the Newmark β scheme is used as follows:

$$
0 = r_{\text{tol}} = \frac{S_{\text{D}}}{\beta \Delta t^2} M_{\text{tol}} \left(u_{\text{tol}}^{n+1} - \tilde{u}_{\text{tol}}^{n+1} \right) - f_{\text{tol}}^{\text{ext}} \left(u_{\text{tol}}^{n+1}, t^{n+1} \right) + f_{\text{tol}}^{\text{int}} \left(u_{\text{tol}}^{n+1} \right) \tag{6.45}
$$

where S_{D} is a switch, which is set to 0 for a static equilibrium problem and 1 for a dynamic transient problem. The matrix r_{tol} is called a residual. The displacement and velocity are solved by time integration:

$$
\begin{aligned}
u_{\text{tol}}^{n+1} &= \tilde{u}_{\text{tol}}^{n+1} + \beta \Delta t^2 a_{\text{tol}}^{n+1} \\
\tilde{u}_{\text{tol}}^{n+1} &= u_{\text{tol}}^n + \Delta t v_{\text{tol}}^n + \frac{\Delta t^2}{2} (1 - 2\beta) a_{\text{tol}}^n
\end{aligned} \tag{6.46}
$$

$$
\begin{aligned}
v_{\text{tol}}^{n+1} &= \tilde{v}_{\text{tol}}^{n+1} + \gamma \Delta t a_{\text{tol}}^{n+1} \\
\tilde{v}_{\text{tol}}^{n+1} &= v_{\text{tol}}^n + (1 - \gamma) \Delta t a_{\text{tol}}^n
\end{aligned} \tag{6.47}
$$

in which the displacement, velocity, and acceleration are given by

$$
u_{\text{tol}} = \left\{ \begin{array}{c} u \\ u^* \end{array} \right\}; \quad v_{\text{tol}} = \left\{ \begin{array}{c} v \\ v^* \end{array} \right\}; \quad a_{\text{tol}} = \left\{ \begin{array}{c} a \\ a^* \end{array} \right\} \tag{6.48}
$$

They have assembled original degrees of freedom and enriched degrees of freedom.

Equation (6.45) is a set of nonlinear equations, so the Newton−Raphson iteration is adopted for solving this equation:

$$A\Delta u = -r(u_n, t^{n+1}), \quad u_{n+1} = u_n + \Delta u \tag{6.49}$$

where A is an equivalent tangent stiffness matrix, such that

$$A = \frac{\partial r}{\partial u} = \frac{s_D}{\beta \Delta t^2} M + \frac{\partial f^{\text{int}}}{\partial u} - \frac{\partial f^{\text{ext}}}{\partial u} \quad \text{for} \quad \beta > 0 \tag{6.50}$$

The derivative of internal force f^{int} with respect to displacement u is divided into two parts, which are called the material stiffness matrix K^{mat} and the geometric stiffness matrix K^{geo}:

$$\frac{\partial f^{\text{int}}}{\partial u} = K^{\text{mat}} + K^{\text{geo}} \tag{6.51}$$

in which

$$
\begin{aligned}
K_{IJ}^{\text{mat}} &= \int_{\Omega_0} B_{0I}^{\text{T}}[C^{\text{SE}}] B_{0J} d\Omega_0 \\
K_{IJ}^{\text{geo}} &= I \int_{\Omega_0} B_{0I}^{\text{T}}[S] B_{0J} d\Omega_0
\end{aligned}
\tag{6.52}
$$

6.5.5 Continuum Element Transformed to Shell

Since the enriched shape function is based on the continuum element, the corresponding extended DOFs are located at the slave node of the CB shell. In order to transform them into the extended DOFs located at the master node on the mid-surface of the shell, the original method can be used to find the CB shell. The transformation matrix T is

$$
T = \begin{bmatrix}
1 & 0 & 0 & h^+ \cos(\bar{e}_{I1}, e_1) & -h^+ \cos(\bar{e}_{I2}, e_1) \\
0 & 1 & 0 & h^+ \cos(\bar{e}_{I1}, e_2) & -h^+ \cos(\bar{e}_{I2}, e_2) \\
0 & 0 & 1 & h^+ \cos(\bar{e}_{I1}, e_3) & -h^+ \cos(\bar{e}_{I2}, e_3) \\
1 & 0 & 0 & -h^- \cos(\bar{e}_{I1}, e_1) & h^- \cos(\bar{e}_{I2}, e_1) \\
0 & 1 & 0 & -h^- \cos(\bar{e}_{I1}, e_2) & h^- \cos(\bar{e}_{I2}, e_2) \\
0 & 0 & 1 & -h^- \cos(\bar{e}_{I1}, e_3) & h^- \cos(\bar{e}_{I2}, e_3)
\end{bmatrix}
\tag{6.53}
$$

Thus, the relation of extended DOFs between master node and slave node is

$$
\begin{Bmatrix} u_i^{+*} \\ v_i^{+*} \\ w_i^{+*} \\ u_i^{-*} \\ v_i^{-*} \\ w_i^{-*} \end{Bmatrix} = T \begin{Bmatrix} u_i^* \\ v_i^* \\ w_i^* \\ \alpha_i^* \\ \beta_i^* \end{Bmatrix} \tag{6.54}
$$

Another important role of the transformation matrix is to transform the internal node force, mass matrix, and tangent stiffness matrix:

$$
\begin{aligned}
\overline{f}_I &= T_I^{\mathrm{T}} \begin{Bmatrix} f_{I+} \\ f_{I-} \end{Bmatrix} \\
\overline{M}_{IJ} &= T_I^{\mathrm{T}} M_{IJ} T_J \\
\overline{K}_{IJ} &= T_I^{\mathrm{T}} K_{IJ} T_J
\end{aligned} \tag{6.55}
$$

in which $\overline{f}_I$, $\overline{M}_{IJ}$, and $\overline{K}_{IJ}$ are the internal node force, mass matrix, and tangent stiffness matrix of the master node of the CB shell element respectively; f_{I+} and f_{I-} are the internal node forces corresponding to the top and bottom slave nodes on the fiber; M_{IJ} and K_{IJ} are mass matrix and tangent stiffness matrix of the slave node of the CB shell element respectively. After transformation, the equations of motion are given by

$$
\begin{aligned}
\overline{M}_{ijIJ} \ddot{u}_{jJ} + \overline{f}_{iI}^{\mathrm{int}} &= \overline{f}_{iI}^{\mathrm{ext}} \\
\overline{M}_{ijIJ}^{+} \ddot{a}_{jJ} + \overline{q}_{iI}^{\mathrm{int}} &= \overline{q}_{iI}^{\mathrm{ext}}
\end{aligned} \tag{6.56}
$$

and the residual equation is solved by Newton–Raphson iteration:

$$
0 = \overline{r}_{\mathrm{tol}} = \frac{s_{\mathrm{D}}}{\beta \Delta t^2} \overline{M}_{\mathrm{tol}} \left(\overline{u}_{\mathrm{tol}}^{n+1} - \widetilde{\mathbf{u}}_{\mathrm{tol}}^{n+1} \right) - \overline{f}_{\mathrm{tol}}^{\mathrm{ext}} \left(\widetilde{u}_{\mathrm{tol}}^{n+1}, t^{n+1} \right) + \overline{f}_{\mathrm{tol}}^{\mathrm{int}} \left(\overline{u}_{\mathrm{tol}}^{n+1} \right) \tag{6.57}
$$

The discretized momentum equation (6.56) and time integration scheme (6.57) are the computational solution of the CB shell, which is the final formula in the X-FEM program.

6.6 CRACK PROPAGATION CRITERION

6.6.1 Stress Intensity Factor Computation

The calculation of the stress intensity factors (SIFs) is based on the work of Nikishkov and Atluri (1987), who proposed an equivalent domain integral method for calculating the fracture parameter SIF of a 3D arbitrary crack. This

Extended Finite Element Method

FIGURE 6.17 A local coordinate
system at the crack tip.

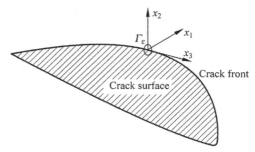

method has been widely used in standard FEM and in this subsection it has
been introduced into the CB shell element of X-FEM.

A local coordinate system (x_1, x_2, x_3) is established at the crack tip as
illustrated in Figure 6.17, where x_1 is located on the crack surface perpendicular to the crack tip; x_3 is located on the crack surface at a tangent to the
crack tip; x_1 and x_2 are perpendicular to x_3. Using these local coordinates, the J
integral corresponding to mode I and mode II cracks, as well as the energy
release rate corresponding to a mode III crack G_{III} (Nikishkov and Atluri,
1987), can be respectively defined as

$$J_k = \lim_{\varepsilon \to 0} \int_{\Gamma_\varepsilon} \left[wn_k - \sigma_{ij}\frac{\partial u_i}{\partial x_k}n_j \right] d\Gamma \qquad (6.58)$$

$$G_{III} = \lim_{\varepsilon \to 0} \int_{\Gamma_\varepsilon} \left[w^{III}n_1 - \sigma_{3j}\frac{\partial u_3}{\partial x_1}n_j \right] d\Gamma \qquad (6.59)$$

where Γ_ε is a closed loop including the crack tip, n_k is a unit normal vector at
some point in Γ_ε, w is strain energy density, and w^{III} is the strain energy
density relating only to ε_{3j} and σ_{3j}.

The contour integral equations (6.58) and (6.59) are often translated into
a domain integral in practical use. As shown in Figure 6.18, V and V_ε are two

FIGURE 6.18 Local coordinates at the crack tip.

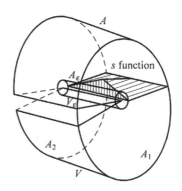

volumes enclosing the crack front; A and A_ε are the side surfaces of these two volumes, and A_1 and A_2 are their bottom surfaces. The function $s = s(x_1, x_2, x_3)$ is defined as zero on A and non-zero on A_ε. The projection area f is on the side surface of A_ε. Thus, Eqs. (6.58) and (6.59) can be transformed into

$$J_k f = - \int_{A-A_\varepsilon} \left[w n_k - \sigma_{ij} \frac{\partial u_i}{\partial x_k} n_j \right] s dA \tag{6.60}$$

$$G_{III} f = - \int_{A-A_\varepsilon} \left[w^{III} n_1 - \sigma_{3j} \frac{\partial u_3}{\partial x_1} n_j \right] s dA \tag{6.61}$$

After the divergence theorem is employed, the above two equations can be transformed into the volume integral forms:

$$J_k f = - \int_{V-V_\varepsilon} \left[w \frac{\partial s}{\partial x_k} - \sigma_{ij} \frac{\partial u_i}{\partial x_k} \frac{\partial s}{\partial x_j} \right] dv - \int_{V-V_\varepsilon} \left[\frac{\partial w}{\partial x_k} - \frac{\partial}{\partial x_j} \left(\sigma_{ij} \frac{\partial u_i}{\partial x_k} \right) \right] s dv$$
$$+ \int_{A_1+A_2} \left[w n_k - \sigma_{ij} \frac{\partial u_i}{\partial x_k} n_j \right] s dA \tag{6.62}$$

$$G_{III} f = - \int_{V-V_\varepsilon} \left[w^{III} \frac{\partial s}{\partial x_k} - \sigma_{3j} \frac{\partial u_3}{\partial x_1} \frac{\partial s}{\partial x_j} \right] dv - \int_{V-V_\varepsilon} \sigma_{ij} \frac{\partial \varepsilon_{ij}^t}{\partial x_1} s dv$$
$$+ \int_{A_1+A_2} \left[w^{III} n_1 - \sigma_{3j} \frac{\partial u_3}{\partial x_1} n_j \right] s dA \tag{6.63}$$

It can be proved that the second terms in Eqs. (6.62) and (6.63) vanish to zero in the case of linear elasticity and zero body force, and it is obvious that the third terms also vanish to zero if the function s takes the value of zero on A_1 and A_2 respectively.

To decouple the energy release rates from J integrals corresponding to the different modes of cracks, a decomposition technique is used to separate the displacement field and stress field (Shivakumar and Raju, 1992). The displacement and stress fields at a point P near the crack tip can be decomposed into symmetrical and asymmetrical parts:

$$\boldsymbol{u} = \boldsymbol{u}^{I} + \boldsymbol{u}^{II} + \boldsymbol{u}^{III} = \left\{ \begin{array}{c} u_1^P + u_1^{P'} \\ u_2^P - u_2^{P'} \\ u_3^P + u_3^{P'} \end{array} \right\} + \left\{ \begin{array}{c} u_1^P - u_1^{P'} \\ u_2^P + u_2^{P'} \\ 0 \end{array} \right\} + \left\{ \begin{array}{c} 0 \\ 0 \\ u_3^P - u_3^{P'} \end{array} \right\} \tag{6.64}$$

$$\boldsymbol{\sigma} = \boldsymbol{\sigma}^{\mathrm{I}} + \boldsymbol{\sigma}^{\mathrm{II}} + \boldsymbol{\sigma}^{\mathrm{III}} = \left\{\begin{array}{c} \sigma_{11}^{P} + \sigma_{11}^{P'} \\ \sigma_{22}^{P} + \sigma_{22}^{P'} \\ \sigma_{33}^{P} + \sigma_{33}^{P'} \\ \sigma_{12}^{P} - \sigma_{12}^{P'} \\ \sigma_{23}^{P} - \sigma_{23}^{P'} \\ \sigma_{33}^{P} + \sigma_{33}^{P'} \end{array}\right\} + \left\{\begin{array}{c} \sigma_{11}^{P} - \sigma_{11}^{P'} \\ \sigma_{22}^{P} - \sigma_{22}^{P'} \\ 0 \\ \sigma_{12}^{P} + \sigma_{12}^{P'} \\ 0 \\ 0 \end{array}\right\} + \left\{\begin{array}{c} 0 \\ 0 \\ \sigma_{33}^{P} - \sigma_{33}^{P'} \\ 0 \\ \sigma_{23}^{P} + \sigma_{23}^{P'} \\ \sigma_{33}^{P} - \sigma_{33}^{P'} \end{array}\right\}$$

$$\tag{6.65}$$

where

$$u_i^{P'}(x_1, x_2, x_3) = u_i^{P}(x_1, -x_2, x_3), \quad \sigma_i^{P'}(x_1, x_2, x_3) = \sigma_i^{P}(x_1, -x_2, x_3) \tag{6.66}$$

Then, the J integral can be decomposed into three parts:

$$J_1 = G_{\mathrm{I}} + G_{\mathrm{II}} + G_{\mathrm{III}} \tag{6.67}$$

where the energy release rates G_{I}, G_{II}, and G_{III} are integrated using the corresponding decomposed displacement and stress fields:

$$G_{\mathrm{I}} = \lim_{\varepsilon \to 0} \int_{\Gamma_\varepsilon} \left[w^{\mathrm{I}} n_k - \sigma_{ij}^{\mathrm{I}} \frac{\partial u_i^{\mathrm{I}}}{\partial x_k} n_j \right] \mathrm{d}\Gamma$$

$$G_{\mathrm{II}} = \lim_{\varepsilon \to 0} \int_{\Gamma_\varepsilon} \left[w^{\mathrm{II}} n_k - \sigma_{ij}^{\mathrm{II}} \frac{\partial u_i^{\mathrm{II}}}{\partial x_k} n_j \right] \mathrm{d}\Gamma \tag{6.68}$$

$$G_{\mathrm{III}} = \lim_{\varepsilon \to 0} \int_{\Gamma_\varepsilon} \left[w^{\mathrm{III}} n_k - \sigma_{ij}^{\mathrm{III}} \frac{\partial u_i^{\mathrm{III}}}{\partial x_k} n_j \right] \mathrm{d}\Gamma$$

As soon as G_{I}, G_{II}, and G_{III} are integrated, the stress intensity factors can be calculated using the formulae:

$$\begin{aligned} K_{\mathrm{I}} &= \sqrt{E^* G_{\mathrm{I}}} \\ K_{\mathrm{II}} &= \sqrt{E^* G_{\mathrm{II}}} \\ K_{\mathrm{III}} &= \sqrt{2\mu G_{\mathrm{III}}} \end{aligned} \tag{6.69}$$

where

$$E^* = \begin{cases} E/(1-\nu)^2 & \text{plane strain} \\ E & \text{plane stress} \end{cases} \tag{6.70}$$

In the calculation, the integral equation (6.68) is implemented in the area containing and adjacent to the crack tip, as shown in Figure 6.19. Here, (x_1, x_2, x_3) is the local coordinate system defined previously. In order to define the s function that is nonzero at the crack tip and zero on the boundary (thick lines in Figure 6.19), s can be defined as

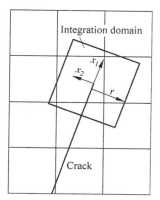

FIGURE 6.19   **Integration domain at the crack tip.**

$$s = \left(1 - \frac{|x_1|}{r}\right)\left(1 - \frac{|x_2|}{r}\right)\left(1 - \frac{|x_3|}{h/2}\right) \tag{6.71}$$

where r is the distance between the crack tip and the boundary, and h is the shell thickness. In this way, the third terms in Eqs. (6.62) and (6.63) vanish to zero.

It must be pointed out that the SIFs of cracks in plates and shells vary along the thickness according to the classical theory of plates and shells. It is neglected here for simplicity and the SIF calculated can be considered as an average value over the shell thickness.

6.6.2 Maximum Energy Release Rate Criterion

We need to find a criterion to evaluate crack propagation or arrest. The crack propagation criterion is based on the work of Chang et al. (2006), who proposed the maximum energy release rate criterion for 3D complex crack problems. From the energy balance theory of Griffith, the stress intensity factor is introduced into this criterion. Not only can the required loading condition be obtained for crack propagation using this criterion, but also the crack propagation direction. Under the circumstances of plane complex fracture or pure mode I, mode II fracture, this criterion may degenerate to the maximum energy release rate criterion for 2D problems. For the 3D complex fracture situation, the computational results of Chang et al. (2006) according to this criterion are compared with the experimental data, and show good agreement. This criterion is described in the following subsection.

The crack propagation criterion should have the general expression:

$$f(K_\mathrm{I}, K_\mathrm{II}, K_\mathrm{III}) = C$$
$$\theta_\mathrm{f} = \Theta(K_\mathrm{I}, K_\mathrm{II}, K_\mathrm{III}) \tag{6.72}$$

in which it provides the conditions for crack propagation and crack direction, where C is a constant related to material properties and θ_f is the crack growth angle.

Polar coordinates are set up at the crack tip, for general complex fracture including mode I, mode II and mode III cracks, and the stress field at the crack tip can be expressed using the stress intensity factors:

$$
\begin{aligned}
\sigma_{rr} &= \frac{1}{\sqrt{2\pi r}} \left[\frac{K_I}{4} \left(5 \cos\frac{\theta}{2} - \cos\frac{3\theta}{2} \right) + \frac{K_{II}}{4} \left(-5 \sin\frac{\theta}{2} + 3 \sin\frac{3\theta}{2} \right) \right] \\
\sigma_{\theta\theta} &= \frac{1}{\sqrt{2\pi r}} \left[\frac{K_I}{4} \left(3 \cos\frac{\theta}{2} + \cos\frac{3\theta}{2} \right) + \frac{K_{II}}{4} \left(-3 \sin\frac{\theta}{2} - 3 \sin\frac{3\theta}{2} \right) \right] \\
\sigma_{r\theta} &= \frac{1}{\sqrt{2\pi r}} \left[\frac{K_I}{4} \left(\sin\frac{\theta}{2} + \sin\frac{3\theta}{2} \right) + \frac{K_{II}}{4} \left(\cos\frac{\theta}{2} + 3 \cos\frac{3\theta}{2} \right) \right] \\
\sigma_{\theta z} &= \frac{K_{III}}{\sqrt{2\pi r}} \cos\frac{\theta}{2} \\
\sigma_{rz} &= \frac{K_{III}}{\sqrt{2\pi r}} \sin\frac{\theta}{2}
\end{aligned}
\tag{6.73}
$$

When a crack propagates a unit length, the energy release rate can be written as

$$
G = \frac{1}{E'}\left(K_I^2 + K_{II}^2 \right) + \frac{1}{2\mu}K_{III}^2, \quad E' = \begin{cases} E & \text{plane stress} \\ E/(1 - \nu^2) & \text{plane strain} \end{cases}
\tag{6.74}
$$

Based on the work of Hussain et al. (1974), Nuismer (1975), Palaniswamy and Knauss (1978), as well as Wu (1978), the maximum energy release rate is established. The idea is that the crack propagates when the energy release rate reaches a critical value. The crack propagation direction is along the orientation that makes the energy release rate attain a maximum value. The expressions are given as

$$
G_{\theta\max} \geq G_C, \quad \theta_f = \theta|_{G_\theta = G_{\theta\max}}
\tag{6.75}
$$

where G_C is material toughness and θ_f is the angle between the propagation direction and the crack tip.

If a crack propagates a distance Δa and kinks into angle θ, which is an angle between the propagation direction and crack tip, as shown in Figure 6.20, the energy release rate can be written as a function of θ:

$$
\begin{aligned}
G_\theta = \lim_{\Delta a \to 0} \frac{1}{2\Delta a} \int_0^{\Delta a} \frac{1}{2} [\sigma_{\theta\theta}(a)\delta_\theta(a + \Delta a) + \sigma_{r\theta}(a)\delta_r(a + \Delta a) \\
+ \sigma_{\theta z}(a)\delta_z(a + \Delta a)]d\xi
\end{aligned}
\tag{6.76}
$$

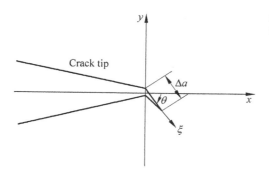

FIGURE 6.20   The crack propagation angle.

in which $\sigma_{ij}(a)$ is the stress field at the crack tip before crack growth, and δ_k is the opening displacement of the crack surface in the k direction.

By substituting Eq. (6.76) into Eq. (6.75), the maximum energy release rate criterion is obtained. However, it is difficult to find the analytical solution of G_θ, which also hinders the application of this criterion to 3D problems. Chang et al. (2006) circumvented solving the cumbersome Eq. (6.76) and assumed that the energy release rate G_θ in the θ direction for a crack under mixed K_{I}, K_{II}, K_{III} states can be approximately assumed to be the same as the energy release rate of the crack under K_{Ieff}, K_{IIeff}, K_{IIIeff} with propagation in its extension direction. The equivalent stress intensity factors K_{Ieff}, K_{IIeff}, K_{IIIeff} can be defined as

$$K_{\mathrm{Ieff}} = \left[\frac{K_{\mathrm{I}}}{4}\left(3\cos\frac{\theta}{2} + \cos\frac{3}{2}\theta\right) + \frac{K_{\mathrm{II}}}{4}\left(-3\sin\frac{\theta}{2} - 3\sin\frac{3}{2}\theta\right)\right]$$

$$K_{\mathrm{IIeff}} = \left[\frac{K_{\mathrm{I}}}{4}\left(\sin\frac{\theta}{2} + \sin\frac{3}{2}\theta\right) + \frac{K_{\mathrm{II}}}{4}\left(\cos\frac{\theta}{2} + 3\cos\frac{3}{2}\theta\right)\right] \quad (6.77)$$

$$K_{\mathrm{IIIeff}} = K_{\mathrm{III}}\cos\frac{\theta}{2}$$

Then, the stress field at the crack tip is

$$\sigma_{\theta\theta} = \frac{K_{\mathrm{Ieff}}}{\sqrt{2\pi r}}, \quad \sigma_{r\theta} = \frac{K_{\mathrm{IIeff}}}{\sqrt{2\pi r}}, \quad \sigma_{\theta z} = \frac{K_{\mathrm{IIIeff}}}{\sqrt{2\pi r}} \quad (6.78)$$

and the approximate expression for G_θ is

$$G(\theta) \approx \frac{\kappa+1}{8\mu}\left[K_{\mathrm{Ieff}}^2 + K_{\mathrm{IIeff}}^2\right] + \frac{1}{2\mu}K_{\mathrm{IIIeff}}^2$$

$$= \frac{1}{2\mu}\cos^2\frac{\theta}{2}\left\{\frac{\kappa+1}{8}\left[K_{\mathrm{I}}^2(1+\cos\theta) - 4K_{\mathrm{I}}K_{\mathrm{II}}\sin\theta + K_{\mathrm{II}}^2(5-3\cos\theta)\right] + K_{\mathrm{III}}^2\right\}$$

$$(6.79)$$

where κ is a material constant: for plane stress $\kappa = (3-\nu)/(1+\nu)$ and for plane strain $\kappa = 3 - 4\nu$.

Substituting the above equation into Eq. (6.75), the criterion of maximum energy release rate for 3D problems can be expressed as

$$\Theta = \frac{\kappa+1}{8}\cos^2\psi\left[2\sin\left(\frac{3}{2}\theta_f+2\gamma\right)+(1+4\sin^2\gamma)\sin\frac{\theta_f}{2}-\sin\frac{3}{2}\theta_f\right]$$

$$+\sin^2\psi\sin\frac{\theta_f}{2}=0$$

$$K_{\text{eff}}^2 = \frac{4\left(K_I^2+K_{II}^2+K_{III}^2\right)}{\kappa+1}\cos^2\frac{\theta_f}{2}$$

$$\left\{\frac{\kappa+1}{8}\cos^2\psi\left[3-4\sin\frac{\theta_f}{2}\sin\left(\frac{\theta_f}{2}+2\gamma\right)-\cos\theta_f\right]+\sin^2\psi\right\}=K_{IC}^2 \quad (6.80)$$

where

$$\gamma = \arctan\left(\frac{K_{II}}{K_I}\right), \quad \psi = \arctan\left(\frac{K_{III}}{\sqrt{K_I^2+K_{II}^2}}\right) \quad (6.81)$$

The first equation in Eq. (6.80) is a nonlinear equation. The propagation angle θ_f can be obtained by solving the equation numerically.

6.7 NUMERICAL EXAMPLES

6.7.1 Mode I Central Through-Crack in a Finite Plate

A plate of size 41 m × 41 m with a central through-crack is subjected to uniform tension at the upper and lower edges, which are parallel to the y-axis, as shown in Figure 6.21. The crack length is $2a$ and the plate thickness is 1 m. Clearly this is a planar mode I fractured plate. It is simulated by the X-FEM program using CB shell elements. The relation between the normalized SIF and the crack length can be obtained from an analytical solution (plane problem), which is plotted along with the numerical results in Figure 6.22.

FIGURE 6.21 Central through-crack in a finite plate.

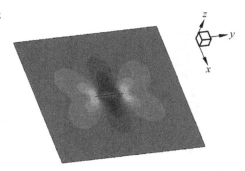

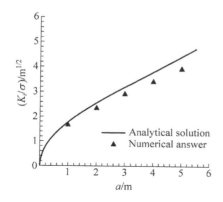

FIGURE 6.22 The relation between the normalized SIF and the crack length.

The numerical result shows reasonable agreement with the analytical solution although there is some discrepancy between them (less than 10%).

6.7.2 Mode III Crack Growth in a Plate

The second example illustrates crack propagation that is dominated by mode III fracture in a plate. The configuration and boundary conditions are shown in Figure 6.23. The size of the plate is 21 m × 21 m, the thickness is 1 m, $\sigma = 10$ MPa, $a = 3.5$ m. The material properties are $E = 210$ GPa, $\nu = 0.3$, and $K_{IC} = 150$ MPa $\bullet$ m$^{1/2}$. The crack position is independent of the mesh, as shown in Figure 6.23. The crack growth path is linear as expected and tearing procession is demonstrated in Figure 6.24.

6.7.3 Steady Crack in a Bending Pipe

The third example is a bending pipe fixed at one end with an embedded crack parallel to the axis at the other end. The crack length is 5.25 m. There is a concentrated force at the point beneath the crack and its direction is upwards

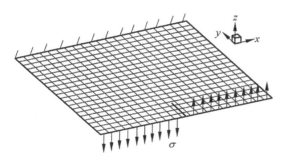

FIGURE 6.23 Mode III fracture in a plate.

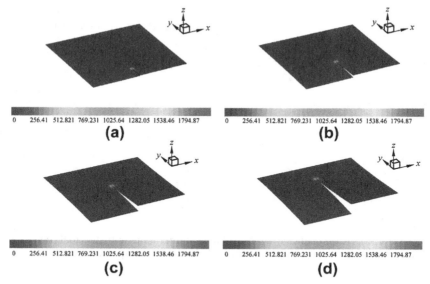

FIGURE 6.24 Crack growing and tearing process (output variable: Mises stress).

with a magnitude of 1653 kN. The pipe dimensions are radius $r = 1.5$ m, length $L = 15$ m, and thickness $h = 0.1$ m. The material parameters are $E = 210$ GPa and $\nu = 0.3$. Figure 6.25 illustrates the mesh and it is clear that the crack does not coincide with the grids. The deforming configuration is shown in Figure 6.26 as the pipe bends upwards and the crack surfaces open. The stress singularity at the crack tip can be found.

FIGURE 6.25 Element mesh of the fractured bending pipe.

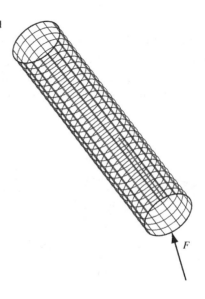

0   8.23529 16.4706   24.7059 32.9412 41.1765 49.4118 57.6471 65.8824

FIGURE 6.26   Deforming configuration of the fractured pipe (output variable: Mises stress).

6.7.4 Crack Propagation Along a Given Path in a Pipe

In this fourth example the dimensions of the pipe and material parameters are the same as those in the third example. The initial length of the crack is 2.25 m and two concentration forces are applied on both surfaces of the crack. The magnitude of these forces is 1653 kN. Because of the geometrical and loading symmetry, a straight growth path is seen. The mode I crack propagates along this path, as shown in Figure 6.27.

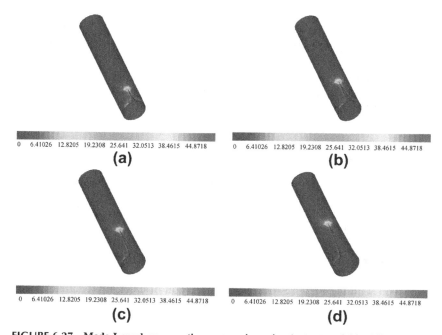

0   6.41026 12.8205 19.2308 25.641 32.0513 38.4615 44.8718
(a)

0   6.41026 12.8205 19.2308 25.641 32.0513 38.4615 44.8718
(b)

0   6.41026 12.8205 19.2308 25.641 32.0513 38.4615 44.8718
(c)

0   6.41026 12.8205 19.2308 25.641 32.0513 38.4615 44.8718
(d)

FIGURE 6.27   Mode I crack propagation process in a pipe (output variable: Mises stress).

FIGURE 6.28 Configuration and mesh of the pipe loaded by an inclined concentration force on the left side of the crack.

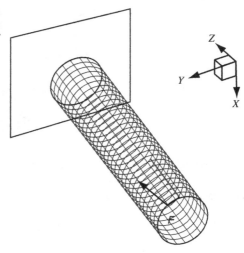

6.7.5 Arbitrary Crack Growth in a Pipe

The fifth example is a pipe fixed at one end with $L = 15$ m, $r = 1.5$ m, and $t = 0.1$ m. There is a crack with an initial length of 4.25 m at the other end. An inclined concentration force of 2×10^4 kN is acting on the left side of the crack. Therefore, a complicated mixed-mode fracture develops in the pipe. The material properties are $E = 210$ GPa, $\nu = 0.3$, and $K_{IC} = 150$ MPa $\bullet$ m$^{1/2}$. The configuration and mesh are illustrated in Figure 6.28.

The computed crack propagation path is shown in Figure 6.29. The crack path is not a straight line but deflects to the loading side because of the asymmetric loading condition. It exhibits a complicated zigzag configuration due to the mixed mode nature of the fracture.

6.8 SUMMARY

Based on the continuum-based (CB) shell element, an X-FEM shell element formula is developed in this chapter. The fracture shell theory and numerical method are also described. Currently, the fracture mechanics of plates and shells are usually studied using two theories: Kirchhoff–Love and Mindlin–Reissner plate and shell bending fracture theories. The former is suitable for thin plates and shells, while the latter is suitable for thicker plates and shells. Comparing the results, it is seen that the solutions of Reissner theory agree well with the FEM results in a small region of crack tip location, while for a large area outside of this small region, the solutions of Kirchhoff theory agree well with the FEM results. The main reason for this arises from the ratio between the thickness of the plate and the computational radius of the crack tip

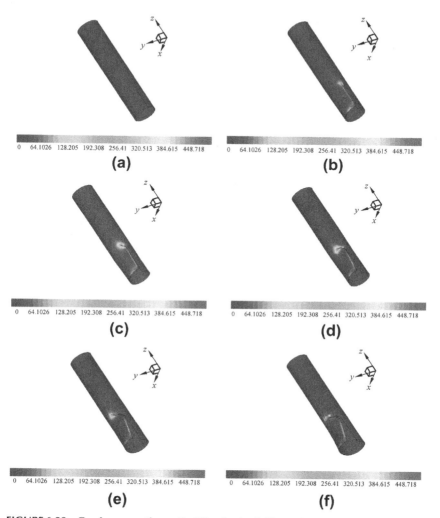

FIGURE 6.29   Crack propagation path of the pipe loaded by an inclined concentration force on the left side of the crack (output variable: Mises stress).

location. The former tends to the thicker plate and the latter tends to the thin plate. Both have their advantages and drawbacks.

Using the method of numerical computation to solve fractured plate and shell problems, the shell element formula for X-FEM was first established by Belytschko. However, because one-point integration was adopted, the hourglass mode may occur in the element, which is an obvious shortcoming. Hourglass control is applied in practice. Another shortcoming of this element is that the fiber elongation is restricted. The thickness of the shell does not change during deformation.

Based on CB shell theory and X-FEM, a unity shell fracture theory for thin and thicker plates and shells was developed by the authors (Zhuang and Cheng, 2011c) and is also described in this chapter. In this work, each CB shell element has eight master nodes on the mid-surface based on shell elements and has 16 slave nodes based on solid elements. By introducing the enriched shape functions into the CB shell element, 3D arbitrary crack growth in curved shells and pipes can be simulated. The crack is independent of the mesh and thus no remeshing is required during crack propagation. The crack is not necessarily perpendicular to the mid-surface for general consideration. The thickness of the shell can change during deformation. A methodology for calculating the stress intensity factors of 3D cracks is adopted and the maximum energy release rate is used as a propagation criterion. Numerical examples prove the capability of this method to capture the path of arbitrary crack growth in shells and pipes.

Chapter 7

Subinterfacial Crack Growth in Bimaterials

Chapter Outline

7.1 **Introduction**   **143**

7.2 **Theoretical Solutions of Subinterfacial Fracture**   **144**

    7.2.1 Complex Variable Function Solution for Subinterfacial Cracks   144

    7.2.2 Solution Considering the Crack Surface Affected Area   147

    7.2.3 Analytical Solution of a Finite Dimension Structure   149

7.3 **Simulation of Subinterfacial Cracks Based On X-FEM**   **153**

    7.3.1 Experiments on Subinterfacial Crack Growth   153

    7.3.2 X-FEM Simulation of Subinterfacial Crack Growth   155

7.4 **Equilibrium State of Subinterfacial Mode I Cracks**   **158**

    7.4.1 Effect on Fracture Mixed Level by Crack Initial Position   158

    7.4.2 Effect on Material Inhomogeneity and Load Asymmetry   159

7.5 **Effect on Subinterfacial Crack Growth from a Tilted Interface**   **163**

7.6 **Summary**   **165**

7.1 INTRODUCTION

Composite materials are widely used in industry, engineering, and civil projects. Their failure receives considerable attention. Fracture problems are studied in this chapter for two types of materials combined with an interface. A crack located at the interface is called an interfacial crack, while a crack located at the area very close to the interface is called a subinterfacial crack. In engineering applications, as well as interfacial cracks, there are many kinds of fracture situations where the crack is not coincident with the interface, but is

Extended Finite Element Method. http://dx.doi.org/10.1016/B978-0-12-407717-1.00007-8

close to it. Thus, these situations are significant in the research on sub-interfacial cracks in bimaterials. Up to now, there has been more research undertaken on interfacial cracks than on subinterfacial cracks. Theoretical research has been ongoing for 50 years, in which the most famous work was that by Rice (1988), who used complex variable functions to find the stress intensity factors of interfacial cracks and the stress and displacement fields near the crack tip. Experimental and computational methods have been adopted to explore the fracture characteristics of static and dynamic interface crack growth in elastic—plastic materials.

Interfacial fracture and subinterfacial fracture in bimaterials have attracted wide interest for their relevance to fracture in composite materials. Sub-interfacial cracks always exhibit a complicated propagation path due to the mixed-mode fracture caused by the combined effects of the interface and loading. Hutchinson et al. (1987) studied the analytical solution for sub-interfacial cracks under far-field dominance. Crack surface contact was considered by Yang and Kim (1992). For cases in which far-field dominance cannot be satisfied, experimental and numerical methods are prevalent. Typical work was carried out by Lee and Krishnaswamy (2000) and Venkatesha et al. (1998). In Lee and Krishnaswamy's work, FEM and auto remeshing are used to simulate crack propagation while the results show some discrepancies with their experimental data. In this chapter, X-FEM is chosen as an alternative to standard FEM and auto remeshing.

Formulae and a program based on the X-FEM algorithm have been developed by the authors, and are applied to simulating subinterfacial crack growth in bimaterials. A brief review on the theoretical research on sub-interfacial cracks is given in section 7.2. Experimental data in finite-dimension specimens and the simulation results of subinterfacial crack growth in bima-terials with X-FEM are included in section 7.3. The simulation results are compared with the experimental data, and show the advantages of X-FEM in dealing with these problems. Then the mechanism of a mode I crack in bimaterials is studied more deeply, with results on the relation between the effects of material inhomogeneity and loading asymmetry being presented in section 7.4. In addition, the effect on crack movement of a tilted interface is discussed in section 7.5.

7.2 THEORETICAL SOLUTIONS OF SUBINTERFACIAL FRACTURE

7.2.1 Complex Variable Function Solution for Subinterfacial Cracks

Hutchinson et al. (1987) studied the analytical solution in the form of a complex variable function to express the stress intensity factor of a sub-interfacial crack under far-field dominance. The starting point is to assume a distance h between the subinterfacial crack and the interface. For a small distance h the crack is very close to the interface, and the singularity solution at the far field of a crack tip must be consistent with the solution for an interfacial crack. Figure 7.1 shows the analytical models corresponding to the interfacial crack, subinterfacial crack, and semi-infinite subinterfacial crack.

The upper and lower parts of an infinite plate are composed of different materials. The shear modulus and Poisson's ratios are G_1, ν_1 and G_2, ν_2 respectively. The crack length is L and the distance from the interface is h, $h \ll L$. A specific point is a distance r from the crack tip. When r is larger than h and less than the geometrical characteristic scale of crack length L, the solution for a subinterfacial crack should be consistent with the solution of the singularity field for an interfacial crack in a semi-infinite plate. Therefore, if we know the solution for an interfacial crack in a semi-infinite plate, the solution for a subinterfacial crack can be deduced.

Related solutions for interfacial cracks have been given by England (1965), Erdogan (1965), and Rice (1988). In a semi-infinite plate including an interfacial crack, the stress expression in front of crack tip ($\theta = 0$) is given by

$$\sigma_{22} + i\sigma_{12} = K(2\pi r)^{-1/2} r^{i\varepsilon} \tag{7.1}$$

where $K = K_1 + iK_2$ is the expression of a complex variable function for the stress intensity factor and

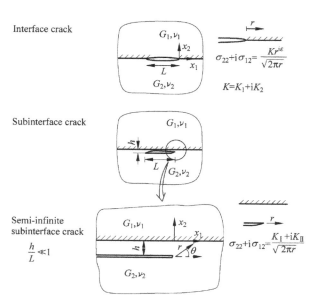

FIGURE 7.1   **Solutions of complex variable function for a subinterfacial crack (Hutchinson et al., 1987).**

$$\varepsilon = \frac{1}{2\pi} \ln \left[\frac{G_1 + G_2(3 - 4\nu_1)}{G_2 + G_1(3 - 4\nu_1)} \right]$$

is a parameter related to the material.

The strain energy release rate of an interfacial crack is

$$G = \frac{(1 - \nu_1)/G_1 + (1 - \nu_2)/G_2}{4 \cosh^2 \pi\varepsilon} K\overline{K} \tag{7.2}$$

in which $\overline{K}$ and K are conjugated.

In the following, a subinterfacial crack is considered. The stress field in front of the crack tip ($\theta = 0$) is given by

$$\sigma'_{22} + i\sigma'_{12} = (K_I + iK_{II})(2\pi r)^{-1/2} \tag{7.3}$$

and the energy release rate is

$$G = \frac{1 - \nu_2}{2G_2} \left(K_I^2 + K_{II}^2 \right) \tag{7.4}$$

As mentioned previously, when $h \ll L$, the far field stresses at the subinterfacial crack tip approximate to the stresses of the interfacial crack solution, i.e., the Williams stress singularity field of an interfacial crack:

$$\sigma_{\alpha\beta} = \mathrm{Re} \left[K(2\pi r)^{-1/2} r^{i\varepsilon} \tilde{\sigma}_{\alpha\beta}(\theta) \right] \tag{7.5}$$

where $\tilde{\sigma}_{\alpha\beta}(\theta)$ are universal regularities of the stress angular distribution, which is only related to the material. Combining Eqs. (7.1), (7.3) and (7.5), and using dimension analysis, the mode I and mode II stress intensity factors of the subinterfacial crack can be obtained:

$$K_I + iK_{II} = cKh^{i\varepsilon} \tag{7.6}$$

On this occasion, considering that the energy release rates for the interfacial crack and subinterfacial crack should also be equal, we can obtain from Eqs. (7.2) and (7.4):

$$K_I^2 + K_{II}^2 = q^2 K\overline{K} \tag{7.7}$$

Combining Eqs. (7.6) and (7.7), we get

$$c = qe^{i\phi} \tag{7.8}$$

where ϕ is a parameter related to the material.

Therefore, the mode I and mode II stress intensity factors of a subinterfacial crack can now be expressed as the corresponding combined stress intensity factor of the interfacial crack ($h = 0$):

$$K_I + iK_{II} = qe^{i\phi} Kh^{i\varepsilon} \tag{7.9}$$

in which the parameters q, ϕ, and ε can be determined by the upper and lower material properties. The expression for ε is given in Eq. (7.1). The parameter $q = [(1 - \beta^2)/(1 + \alpha)]^{1/2}$, and the Dundur parameters α and β, for the plane strain, are given by

$$\alpha = \frac{G_1(1 - \nu_2) - G_2(1 - \nu_1)}{G_1(1 - \nu_2) + G_2(1 - \nu_1)}$$

$$\beta = \frac{1}{2} \frac{G_1(1 - 2\nu_2) - G_2(1 - 2\nu_1)}{G_1(1 - \nu_2) + G_2(1 - \nu_1)}$$

(7.10)

By numerical calculation and fitting, the value of ϕ can be found in terms of α and β:

$$\phi = 0.1584\alpha + 0.0630\beta \tag{7.11}$$

7.2.2 Solution Considering the Crack Surface Affected Area

There is a shortcoming in the analytical solution of subinterfacial cracks deduced by Hutchinson et al., as discussed in subsection 7.2.1, in that it does not consider contact between the crack surfaces. However, in the actual situation, because of the inhomogeneity of the materials on both sides of the crack, it is difficult to prevent contact between crack surfaces, unless the materials are only subjected to uniform tension stresses and have no shear stress. Yang and Kim (1992) considered this point and the problem was described as shown in Figure 7.2.

An infinite plate composed of two kinds of materials, material 1 in the lower part and material 2 in the upper part. A crack is located below the interface at a distance h. A plane coordinate axis is set up so the crack is located on the x-axis. The endpoints of the crack are from -1 to 1. The distance of the plate is subjected to normal stress σ and shear stress τ. The boundary conditions of this problem are given as

$$\text{Far field :} \quad \sigma_{xy} = \tau, \ \sigma_{yy} = \sigma, \ \varepsilon_{xx} = \text{constant} \tag{7.12}$$

$$\text{Interface :} \quad [\sigma_{xy}] = 0, \ [\sigma_{yy}] = 0, \ [u_x] = 0, \ [u_y] = 0 \tag{7.13}$$

$$\text{Crack surface :} \quad \begin{array}{l} \sigma_{xy} = 0, \ \sigma_{yy} = 0, \ [u_y] \geq 0, \ -1 \leq x \leq b \\ \sigma_{xy} = 0, \ \sigma_{yy} < 0, \ [u_y] = 0, \ b \leq x \leq 1 \end{array} \tag{7.14}$$

The square brackets in Eq. (7.13) represent a jump on both sides of the interface. It can be seen from Eq. (7.14) that there is partial contact in the crack length $b \leq x \leq 1$ and the normal stress is stipulated as zero at the end of the contact length:

$$\sigma_{yy}(b, 0) = 0 \qquad (7.15)$$

This problem is too complicated to directly deduce the analytical solution. Thus, a numerical method can be used in the calculation. The crack is represented using the slipping dislocation density $b_x(s)$ distributed on $(-1,1)$ and the opening dislocation density $b_y(s)$ distributed on $(-1,b)$. Their existence has an offsetting effect at the crack location for the boundary condition at the far field. By utilizing the boundary condition (7.12), we have Cauchy singularity integral equations as follows:

$$\int_{-1}^{b} \frac{2b_y(s)}{x-s} ds + \int_{-1}^{b} b_y(s) K_{11}(x,s,h) ds + \int_{-1}^{1} b_x(s) K_{12}(x,s,h) ds = \frac{\sigma}{D}, \quad -1 < x < b$$

$$\int_{-1}^{1} \frac{2b_x(s)}{x-s} ds + \int_{-1}^{b} b_y(s) K_{21}(x,s,h) ds + \int_{-1}^{1} b_x(s) K_{22}(x,s,h) ds = \frac{\tau}{D}, \quad -1 < x < 1$$

$$(7.16)$$

where

$$K_{11}(x,s,h) = \frac{(\lambda+\delta)(x-s)}{(x-s)^2 + 4h^2} - \frac{8\delta h^2 \left[(x-s)^3 - 12(x-s)h^2\right]}{\left[(x-s)^2 + 4h^2\right]^3} + \frac{16h^2\delta(x-s)}{\left[(x-s)^2 + 4h^2\right]^2}$$

$$K_{12}(x,s,h) = -\frac{2(\lambda-\delta)h}{(x-s)^2 + 4h^2} + \frac{8\delta h^2 \left[6(x-s)^2 h - 8h^3\right]}{\left[(x-s)^2 + 4h^2\right]^3}$$

$$K_{21}(x,s,h) = \frac{2(\lambda-\delta)h}{(x-s)^2 + 4h^2} - \frac{8\delta h^2 \left[6(x-s)^2 h - 8h^3\right]}{\left[(x-s)^2 + 4h^2\right]^3}$$

$$K_{22}(x,s,h) = \frac{(\lambda+\delta)(x-s)}{(x-s)^2 + 4h^2} - \frac{8\delta h^2 \left[(x-s)^3 - 12(x-s)h^2\right]}{\left[(x-s)^2 + 4h^2\right]^3} \times \frac{16h^2\delta(x-s)}{\left[(x-s)^2 + 4h^2\right]^2}$$

$$D = \frac{\mu_1}{\pi(\kappa_1 + 1)}; \quad \lambda = \frac{\alpha+\beta}{\beta-1}; \quad \delta = \frac{\beta-\alpha}{\beta+1}$$

$$\alpha = \frac{\mu_1(\kappa_2 + 1) - \mu_2(\kappa_1 + 1)}{\mu_2(\kappa_1 + 1) + \mu_1(\kappa_2 + 1)}; \quad \beta = \frac{\mu_1(\kappa_2 - 1) - \mu_2(\kappa_1 - 1)}{\mu_2(\kappa_1 + 1) + \mu_1(\kappa_2 + 1)}$$

$$(7.17)$$

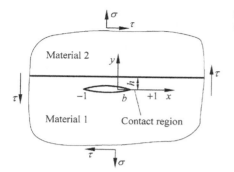

FIGURE 7.2   Crack surface affected area (Yang and Kim, 1992).

In order to guarantee the unique value condition of displacement, the following formulae must be satisfied:

$$\int_{-1}^{1} b_x(s)\mathrm{d}s = 0, \qquad \int_{-1}^{1} b_y(s)\mathrm{d}s = 0 \tag{7.18}$$

This problem cannot be solved by simultaneous equations (7.16) and (7.18) because b is unknown. Using Eq. (7.15), the following function is defined:

$$F(b) = \sigma_{yy}(b,0,h) = \int_{-1}^{b} \frac{2b_y(s)}{b-s}\mathrm{d}s + \int_{-1}^{b} b_y(s)K_{11}(b,s,h)\mathrm{d}s + \int_{-1}^{1} b_s(s)K_{12}(b,s,h)\mathrm{d}s$$

$$\tag{7.19}$$

Therefore, the equations are closed forms. As an iteration method, it is assumed that the value of b and the dislocation densities $b_x(s)$ and $b_y(s)$ are solved using Eqs. (7.16) and (7.18) respectively. Then they are substituted into Eq. (7.19) and we check whether $F(b)$ is equal to zero. If it is not equal to zero, the value of b is modified, and then we recalculate $F(b)$ until it converges.

This solution is more precise than the solution given by Hutchinson et al. A comparison of the two methods is listed in Table 7.1 for solving the stress intensity factors corresponding to tension load and shear load.

7.2.3 Analytical Solution of a Finite Dimension Structure

In the previous discussion, the research focused on the infinite plate. Because of the complexity of the finite dimension problem, experimental and computational methods are usually used in research work. Lee and Krishnaswamy (2000) carried out experiments on subinterfacial crack static growth in bimaterials of finite dimension specimens. In their work, the analytical solution and numerical simulation were also carried out simultaneously. In this

TABLE 7.1 Comparison of Two Methods (Yang and Kim, 1992)

	Tensile stress field			Shear stress field	
	Solution of Hutchinson et al.	Solution of Yang and Kim		Solution of Hutchinson et al.	Solution of Yang and Kim
$K_I/\sqrt{\pi}\sigma$ at the right crack tip	0.8023	0.7978	$K_I/\sqrt{\pi}\tau$ at the left crack tip	0.1253	0.1828
$K_{II}/\sqrt{\pi}\sigma$ at the right crack tip	0.1253	0.1326	$K_{II}/\sqrt{\pi}\tau$ at the left crack tip	0.8023	0.8096

subsection, their analytical results are firstly described; the experimental data are later discussed in detail.

The mechanical model considered by Lee and Krishnaswamy is shown in Figure 7.3. The two materials on either side of the interface are PMMA and Al6061 respectively, which are subjected to three-point quasi-static loading. The distances from the loading direction to the upper and lower sustaining points are L_1 and L_2 respectively. A loading point is located at a geometrical central axis of the specimen. The distance from the interface to the loading point is equal to s. The crack with length a is located at the upward side of the interface, and the distance between them is d. This problem is resolved into two sub-models: (1) a bimaterial specimen without a crack subjected to loading, as shown in Figure 7.3(b); (2) a bimaterial specimen with a crack and loading free, as shown in Figure 7.3(c). Under linear elastic conditions, the

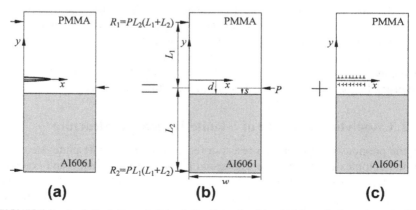

FIGURE 7.3 Analytical research for a finite dimension bimaterial cracked specimen under three-point bending (Lee and Krishnaswamy, 2000).

analytical solutions from these two sub-models are superposed and the solution of the original problem can be obtained.

Clearly, the solution for sub-model (1) is easily solved because it is a bending problem of a Euler–Bernoulli beam. The stress solutions are given by the following equations:

$$\sigma_{yy}(x,d) = \begin{cases} \dfrac{PL_1}{L_1+L_2}\left\{\dfrac{12}{th^3}\right\}[L_2 - s + d]\left[\dfrac{h}{2} - x\right], & s > b \\[3mm] \dfrac{PL_2}{L_1+L_2}\left\{\dfrac{12}{th^3}\right\}[L_1 + s - d]\left[\dfrac{h}{2} - x\right], & s < b \end{cases}$$

$$\sigma_{xy}(x,d) = \begin{cases} -\dfrac{PL_2}{L_1+L_2}\dfrac{1}{th}, & s > b \\[3mm] \dfrac{PL_1}{L_1+L_2}\dfrac{1}{th}, & s < b \end{cases}$$

(7.20)

The solution for sub-model (2) is found using a similar method given by Yang and Kim (1992): the crack is represented by the distributed dislocation density, which offsets the stress solved by sub-model (1) at the crack location. It is difficult to provide the other solution using dislocation to represent the boundary crack; thus, Lee and Krishnaswamy have adopted the dislocation solution of a central crack. Mirror symmetry technology is used for sub-model (2), as illustrated in Figure 7.4.

The disadvantage of this treatment is that it cannot reflect the original problem precisely, because there is a loading-free boundary on the left side of the original specimen for sub-model (2). This condition is not satisfied when it is treated as a central crack. Thus, it is necessary to consider the error resulting from this deviation. Therefore, it is only an approximate treatment method.

The Cauchy singularity integral equations of sub-model (2) are

$$\int_{-a}^{a}\frac{2b_y(s)}{x-s}ds + \int_{-a}^{a}b_y(s)K_{11}(x,s,d)ds + \int_{-a}^{a}b_x(s)K_{12}(x,s,d)ds = -\frac{\sigma_{yy}(s,d)}{D}$$

$$\int_{-a}^{a}\frac{2b_x(s)}{x-s}ds + \int_{-a}^{a}b_y(s)K_{21}(x,s,d)ds + \int_{-a}^{a}b_x(s)K_{22}(x,s,d)ds = -\frac{\sigma_{xy}(s,d)}{D}$$

(7.21)

In order to guarantee the unique value condition of displacement, the following formulae must be satisfied:

$$\int_{-a}^{a}b_x(s)ds = 0, \qquad \int_{-a}^{a}b_y(s)ds = 0$$

(7.22)

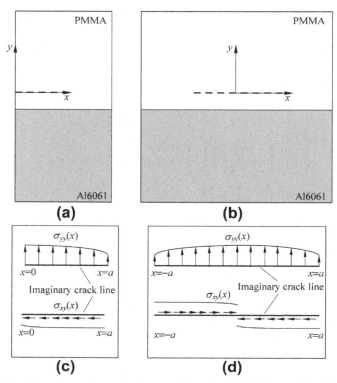

FIGURE 7.4 Mirror symmetry technology used for sub-model (2) (Lee and Krishnaswamy, 2000).

By solving the above two equations, the analytical solution of sub-model (2) can be obtained.

With regard to the stress intensity factor, they can be solved using dislocation density:

$$K_I = -\frac{2\mu_1}{\sqrt{1+\kappa_1}}\sqrt{\pi a}\left[\sqrt{a^2-x^2}b_y(x)\right]_{x\to a}$$

$$K_{II} = \frac{2\mu_1}{\sqrt{1+\kappa_1}}\sqrt{\pi a}\left[\sqrt{a^2-x^2}b_x(x)\right]_{x\to a}$$

(7.23)

In their analytical work, Lee and Krishnaswamy found $K_{II} = 0$ at d/a approximately equal to 0.25, when $w = 150$ mm, $h = 150$ mm, $t = 9$ mm, $L_1 = 100$ mm, $L_2 = 150$ mm, $a = 70$ mm, and $s = 10$ mm. This means that the crack grows along a straight line, which indicates a mode I crack. They also observed this phenomenon experimentally, which is a key point in the numerical studies in this chapter.

7.3 SIMULATION OF SUBINTERFACIAL CRACKS BASED ON X-FEM

7.3.1 Experiments on Subinterfacial Crack Growth

Experiments were conducted by Lee and Krishnaswamy (2000). The effect on mixed-mode crack propagation was studied in detail, taking account of the interface of a bimaterial and the loading. In their experiments, as shown in Figure 7.5, a specimen with an initial crack was subjected to three-point quasi-static increasing loading and the interface was adjacent to the crack. By changing the boundary condition and the distance between the interface and the initial crack, different crack growth paths were observed. For each crack path, the magnitude of the stress intensity factor $|K| = \sqrt{K_I^2 + K_{II}^2}$ and the phase angle $\phi = \arctan(K_{II}/K_I)$ were recorded and plotted. Some parameters of the experiment are: specimen width w, height h and thickness t; distances from the initial crack to the left and right sustaining points L_1 and L_2 respectively; distance between the interface and crack d; distance between the interface and loading point s, and initial crack length a. The experimental configuration is illustrated in Figure 7.5. The two materials on either side of the interface were PMMA ($E = 3.24$ GPa, $v = 0.35$) and Al6061 ($E = 69$ GPa, $v = 0.3$) respectively.

For case I, as shown in Figure 7.6, when $w = 300$ mm, $h = 150$ mm, $t = 9$ mm, $L_1 = L_2 = 150$ mm, $a = 30$ mm, and $d = s = 10$ mm, the crack in PMMA material deflected leftwards from the interface initially and then took an equilibrium path that propagated almost parallel to the interface. $|K|$ was around $1.1-1.2$MPa $\cdot$ m$^{1/2}$, which is the fracture toughness of homogeneous PMMA. The phase angle ϕ increased from $-15°$ to zero and then stayed at zero during the rest period.

For case II, corresponding to the conditions in case I, the left sustaining point moved rightwards by a certain distance, as shown in Figure 7.7. For

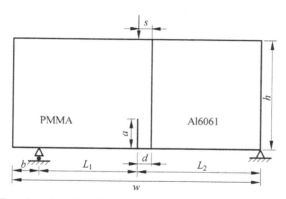

FIGURE 7.5   Experiment configuration.

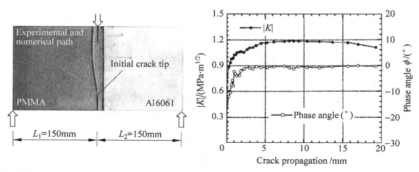

FIGURE 7.6 Experimental data and crack growth configuration for case I (Lee and Krishnaswamy, 2000).

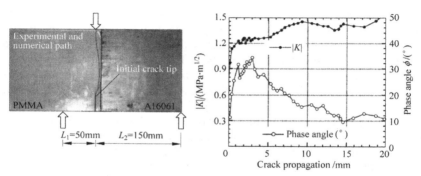

FIGURE 7.7 Experimental data and crack growth configuration for case II (Lee and Krishnaswamy, 2000).

$w = 300$ mm, $h = 150$ mm, $t = 9.5$ mm, $L_1 = 50$ mm, $L_2 = 150$ mm, $a = 30$ mm, and $d = s = 10$ mm, the crack was initially attracted to the interface, then grew along it and finally rebounded from the interface. $|K|$ increased from 1.0 to 1.5MPa $\cdot$ m$^{1/2}$, while the phase angle ϕ decreased from 30° to 10°.

For case III, an interesting problem is considered: now that the crack paths have different configurations in cases I and case II respectively, does an equilibrium state exist in which the crack propagates linearly while the effects of the interface and loading counteract each other? This was achieved by selecting appropriate parameters and sustaining points, as shown in Figure 7.8: $w = 300$ mm, $h = 150$ mm, $t = 9$ mm, $L_1 = 100$ mm, $L_2 = 150$ mm, $a = 70$ mm, $d = 16$mm, and $s = 10$ mm. Here, d/a is approximately equal to 0.23, which is close to the analytical result $d/a = 0.25$.

The experiments of Lee and Krishnaswamy demonstrated the different geometrical and constraint conditions, as well as the effect of the initial crack location on subinterfacial crack growth in bimaterials. The crack growth path was strongly affected by the geometrical condition and material

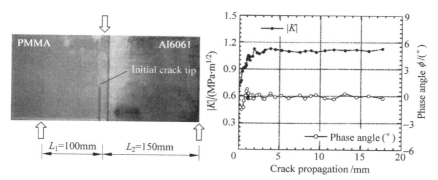

FIGURE 7.8 Experimental data and crack growth configuration for case III (Lee and Krishnaswamy, 2000).

inhomogeneity; the experiment data provided a clear description of this. Meanwhile, the numerical simulation results for the crack path were given by Lee and Krishnaswamy using the finite element method and an automatic mesh technique. The simulation results for cases I and III agreed well with the experimental data. However, for case II, the crack path obtained by numerical computation also firstly propagated toward the interface, but the crack path did not agree well with the experiments (Lee and Krishnaswamy, 2000). At the same time, they indicated that further refinement of the mesh did not improve the simulation results. Perhaps this is a limitation of using the general finite element method to simulate the complexity of the crack path, because the crack must propagate along the element edges in the general finite element method. To remedy these shortcomings, we have developed X-FEM as a numerical simulation tool.

7.3.2 X-FEM Simulation of Subinterfacial Crack Growth

In this subsection, X-FEM is used to capture the actual crack growth path. Instead of steadily increasing loading, a constant loading value is applied to the specimen for simplification. For these three cases, the loading values are 360, 800, and 400 kN respectively. The results for stress intensity factor $|K|$ and phase angle ϕ are also output and plotted compared with the experimental data.

For case I, as shown in Figure 7.9, the numerical crack path grows leftwards from the interface initially and then stays parallel to the interface, which is in accordance with the experimental data, as shown in Figure 7.6. As crack length increases, $|K|$ increases as expected. Since the loading value is different from the experimental value, $|K|$ is much larger than the experimental data, but the phase angle ϕ shows the same tendency in case I, increasing from $-6°$ and staying at zero afterwards (the solid line is the least-square fit of discrete data), as given in Figure 7.9(a). Though the magnitudes of stress intensity factor

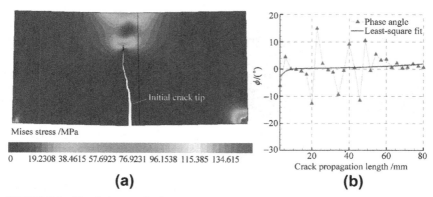

FIGURE 7.9 Simulation results for case I (360 kN).

$|K|$ differ from the experimental data, in the linear elastic situation, the ratio between K_I and K_{II} is insensitive to the magnitude of loading.

For case II, the crack propagates towards the interface initially and rebounds when it gets very close to the interface, as shown in Figure 7.10, then the crack grows along the interface. The whole process is the same as in the experiment. The numerical results provided by Lee and Krishnaswamy show some discrepancies with the experiment: the crack propagates into the interface and does not rebound. This phenomenon is also observed using a coarse mesh in the X-FEM, but better results are obtained when the mesh is refined. The mesh density shown here is 53×105. It is worth noting that the crack path obtained using X-FEM is closer to the experimental data than results using the general FEM. This improvement can be attributed to the X-FEM, which allows an arbitrary crack path not limited by the mesh. The phase angle ϕ decreases from $36°$ to $20°$ when the crack propagation is

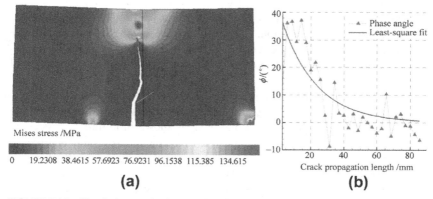

FIGURE 7.10 Simulation results for case II (800 kN).

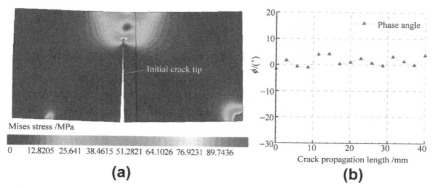

(a)   **(b)**

FIGURE 7.11   Simulation results for case III (400 kN).

20 mm (the final experiment data) and continues to decrease until it reaches zero, which implies a mode I fracture with a straight crack path, as illustrated in Figure 7.10(b).

The case III simulation is much simpler than the former two. The crack propagates along a straight line, as shown in Figure 7.11. The phase angle ϕ remains around zero, which implies a mode I fracture. It is interesting that the interface and loading asymmetry counteract each other in this case. Figure 7.11 shows the results obtained using X-FEM. Both the crack path and the phase angle agree well with the experimental data.

In order to validate the reliability of the computational results, the loading values are modified by 500, 900, and 500 kN, corresponding to the three cases of experiment. The simulation results of phase angle ϕ are plotted in Figure 7.12. They are similar to the results calculated previously, and the tendency of ϕ shows a pleasing agreement with the experimental data.

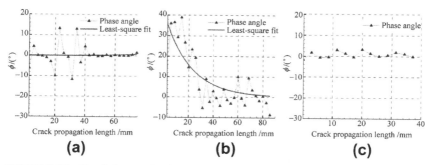

(a)   **(b)**   **(c)**

FIGURE 7.12   Simulation results for three cases of experiments under different loading conditions, 500, 900, and 500 kN.

7.4 EQUILIBRIUM STATE OF SUBINTERFACIAL MODE I CRACKS

7.4.1 Effect on Fracture Mixed Level by Crack Initial Position

In this subsection, a numerical method is used to study the relations between crack position d and length a in the equilibrium state. By keeping the loading condition of case III and changing crack length a, the stress intensity factor for a mode II crack, K_{II}, is calculated corresponding to different crack locations d. The relations between K_{II} and d, and between K_{II} and d/a, are plotted in Figures 7.13 and 7.14 respectively.

Figures 7.13 and 7.14 reveal that changing crack length a can not only increase or decrease K_{II} as, when the crack position d is varied over a large range, the curves representing different crack lengths a intersect at some point. This means that the variation of the mixed level of subinterfacial cracks is very complicated when the crack length and position are changed. Meanwhile, it can be seen that K_{II} is usually less than 10% of K_I during the computational process, although variation in K_{II} may occur. Thus, the influence on crack mixed level from the variation of crack length is very limited.

At $a = 70$ mm, the same situation as the case III experiment, K_{II} is equal to zero when $d^* = 18.75$ mm. Here, $d^*/a = 0.267$, which is close to the analytical solution of $d/a = 0.25$. As the crack length changes, the variations in the crack equilibrium position and relative equilibrium position are shown in Figures 7.15 and 7.16 respectively. When the crack length increases, the crack equilibrium position d^* increases linearly, while the relative equilibrium position d^*/a exhibits a nonlinear decrease.

FIGURE 7.13 Relation between K_{II} and crack position d.

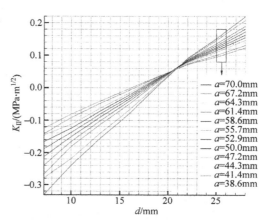

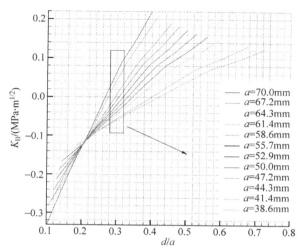

FIGURE 7.14   Relation between K_{II} and relative crack position d/a.

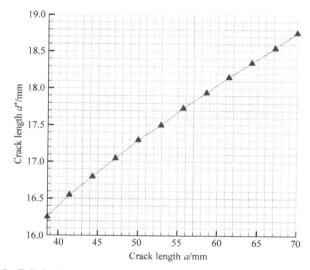

FIGURE 7.15   Relation between d^* and crack length a.

7.4.2 Effect on Material Inhomogeneity and Load Asymmetry

The case III experiment draws much of our curiosity as the crack propagates linearly in an inhomogeneous material as long as the boundary condition is properly designed, which is similar to mode I crack growth in isotropic material under a symmetric load. This means that there exists an 'equilibrium state' between material inhomogeneity and loading asymmetry. To prove this

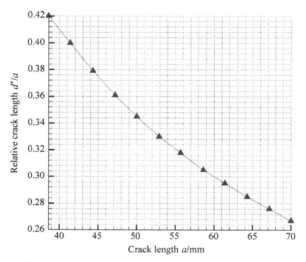

FIGURE 7.16   **Relation between *d*/a* and crack length *a*.**

point, two situations are simulated: (1) PMMA–Al6061 bimaterials with a symmetrical boundary condition (propped on both ends of the specimen); (2) the same boundary condition as in case III was used with homogeneous material (Al6061). The distances between the crack tip and mid-axis of the specimen (equal to 6 mm at the initial time) during crack propagation are recorded and plotted in Figure 7.17.

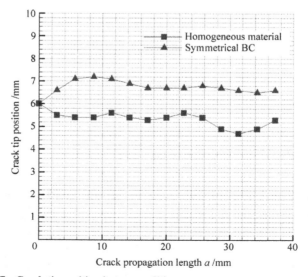

FIGURE 7.17   **Crack tip position in two conditions.**

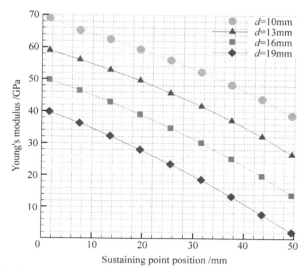

FIGURE 7.18   Relations between boundary condition and material property.

In these two conditions, cracks are inclined in different directions. When the boundary condition is symmetrical, the crack propagates outwards from the mid-axis, due to the interface. In the other case with homogeneous material, the crack propagates towards the mid-axis, which is due to the asymmetrical boundary condition. This proves our conjecture that it is because the effects of material inhomogeneity and loading asymmetry counteract each other that a mode I crack is observed in the case III experiment. Further research has been carried out to reveal this relationship. The distance between the left sustaining point and the left endpoint of the specimen was changed (from 0 to 50 mm) and the Young's modulus of the left material was simultaneously changed to get mode I crack growth (the phase angle changes from $-0.1°$ to $0.1°$ during crack propagation). The relations between the boundary condition and the material property are plotted in Figure 7.18.

It can be seen from Figure 7.18 that Young's module decreases monotonously with increasing distance between the sustaining point and the left endpoint of the specimen, which can be easily understood by the fact that they have opposite effects on the direction of crack propagation. It is worthwhile pointing out that when the distance is 50 mm, which is the same value as in case III, Young's modulus of the left material in the equilibrium state is about 14 GPa. This is different from Young's modulus of 3.24 GPa of PMMA in the experiment. It is caused by a more restricted criterion of the equilibrium state we have chosen in the computation: the phase angle must be limited between $-0.1°$ and $0.1°$ and K_{II} is less than 0.2% of K_I during propagation. This means that the stress intensity factor (SIF) for mode II fracture is less than 2% of the SIF of mode I fracture, hence a more perfect mode I crack is observed.

When the subinterfacial crack shows mode I propagation, the relation between Young's modulus of the left side material and the left sustaining point is illustrated in Figure 7.18. There are four curves corresponding to the lengths 10, 13, 16, and 19 mm, which are the distances from the initial crack to the interface of bimaterials. All four curves show a tendency to decrease monotonously. This can be explained by an opposite effect on mode I fracture of material inhomogeneity and load asymmetry when the initial crack position is determined. Namely, the larger the difference of Young's modulus on both sides of the materials, the more distance offset is seen from the left sustaining point to the left endpoint in order to make the subinterfacial crack realize mode I growth. It can also be seen in Figure 7.18 that the curves appear to move with a downwards translational tendency with increasing distance between the initial crack position and the interface. This can be explained since the greater the distance offset from the initial crack position to the interface, under the same sustaining points condition, the lower the Young's modulus required on the left side material in order to realize mode I growth, which results in more distinct material inhomogeneity; under the same material conditions, the sustaining point is much closer to the left endpoint, which results in no obvious loading asymmetry. In addition, when the initial crack positions increase in an arithmetic progression, the curves show a downwards translational trend over the same intervals. This can be explained by there being a linear effect on mode I crack growth from the initial crack position.

If the variables are transformed into dimensionless forms and their relations are fitted using a polynomial, under the conditions of $E_2 = 69$ GPa, $L = 150$ mm, $a = 70$ mm, the relations of material inhomogeneity, loading asymmetry, and initial crack position can be expressed by the following experience equation:

$$\left(\frac{E_1}{E_2}\right) = -1.5229\left(\frac{b}{L}\right)^2 - 1.0152\left(\frac{b}{L}\right) + 3.227\left(\frac{d}{a}\right) + 0.1272 \quad (7.24)$$

where E_1 and E_2 are Young's moduli of the left and right side materials respectively; b is a distance from the left sustaining point to the left end of the specimen; $L = b + L_1$ is the half-length of the specimen; d is the distance between the initial crack and interface; a is the initial crack length.

Subinterfacial mode I crack growth can be realized in bimaterials if Eq. (7.24) is satisfied. The significance of this experience equation is that it can guide experimental research and engineering applications. There is a linear effect on material inhomogeneity from the initial crack position, while there is a nonlinear effect from the loading asymmetry. Comparitive results calculated by Eq. (7.24) are given in Figure 7.19.

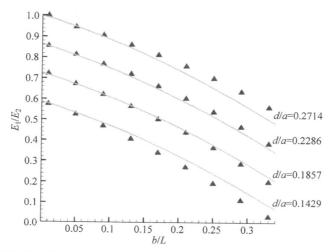

FIGURE 7.19   Equilibrium state mode I crack growth corresponding to relations of material inhomogeneity, loading asymmetry, and initial crack position.

7.5 EFFECT ON SUBINTERFACIAL CRACK GROWTH FROM A TILTED INTERFACE

In the previous simulations, both the interface of bimaterials and the direction of the initial crack are upright. In this section, a tilted material interface is considered. Under the conditions of the case I experiment, the angles between the interface and the upright direction are changed in the range of $-10°$ and $10°$. The crack growth paths corresponding to various angles are plotted in Figures 7.20–7.23.

The crack growth paths corresponding to various angles are plotted in Figure 7.24. It can be seen that the larger the angle between the interface and

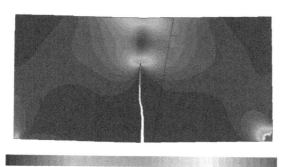

0   19.2308 38.4615 57.6923 76.9231 96.1538 115.385 134.615

FIGURE 7.20   Angle between the interface and upright direction of $-10°$.

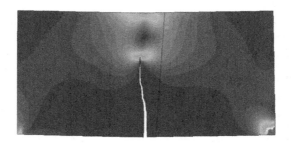

0   19.2308 38.4615 57.6923 76.9231 96.1538 115.385 134.615

FIGURE 7.21   Angle between the interface and upright direction of −5°.

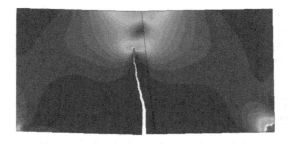

0   19.2308 38.4615 57.6923 76.9231 96.1538 115.385 134.615

FIGURE 7.22   Angle between the interface and upright direction of 5°.

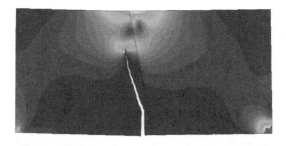

0   19.2308 38.4615 57.6923 76.9231 96.1538 115.385 134.615

FIGURE 7.23   Angle between the interface and upright direction of 10°.

upright direction, the more leftward the crack propagates. This shows that the relatively harder material interface in bimaterials has a clear pushing effect on the growth direction of subinterfacial cracks. In this example, the modulus of the right material (AL6061) is much harder than that of the left material

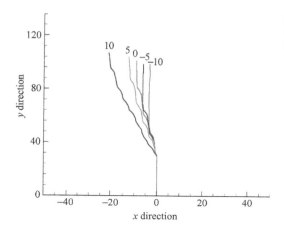

FIGURE   7.24   Subinterfacial crack growth paths corresponding to various interface angles.

(PMMA). This pushing effect is sometimes referred to as "bully the weak and fear the strong".

7.6 SUMMARY

The X-FEM algorithm is discussed in this chapter for simulating subinterfacial crack growth in bimaterials. Because of the complexity of the problem, theoretical research is usually confined to the semi-infinite plane problem. This may improve the precision of the theoretical solution if crack surface contact is considered. The analytical solution for a finite dimension specimen can be obtained by superposition of the solutions for a single dislocation. Experimental study has revealed that the subinterfacial crack may propagate along different paths in the PMMA/Al6061 specimen under different boundary conditions. However, there exists an equilibrium position for crack growth along a straight line. Namely, the subinterfacial crack realizes mode I propagation in bimaterials.

Numerical study was carried out using the X-FEM program developed by the authors to simulate the experiments of Lee and Krishnaswamy (2000). The simulation results of crack growth paths and phase angles agreed well with the experimental data. Because of the advantage of X-FEM compared with conventional FEM, the complex paths of crack growth can be captured precisely. The drawback of crack propagation being dependent on the element boundary in conventional FEM is avoided.

Under the conditions of the case III experiment, the numerical computation reveals that the mode II stress intensity factor K_{II} is equal to zero when the relative equilibrium position $d^*/a = 0.267$, which is close to the analytical solution of $d/a = 0.25$. As the crack length increases, the crack equilibrium position d^* increases linearly, while the relative equilibrium position d^*/a tends to decrease in nonlinear fashion.

The numerical computation results indicate that material inhomogeneity and loading asymmetry in bimaterial specimens have opposite effects on the direction of subinterfacial crack growth. When their effects counteract each other, the subinterfacial crack realizes mode I propagation in bimaterials. For different initial crack positions, further numerical research provides the equilibrium curves corresponding to the material elastic moduli and sustaining point locations. This quantitative relationship is revealed and also fitted by using dimensionless polynomial forms. Subinterfacial crack mode I growth can be realized in bimaterials if the experience equation (7.24) is satisfied; this can be used to guide experimental research. There is a linear effect on material inhomogeneity from the initial crack position, while there is a nonlinear effect from loading asymmetry.

In addition, when the angles between the interface and upright direction are changed, subinterfacial crack propagation is deflected to the soft material side. This shows that the relatively harder material has an obvious pushing effect on the growth direction of subinterfacial cracks. This pushing effect is sometimes referred to as "bully the weak and fear the strong".

Chapter 8

X-FEM Modeling of Polymer Matrix Particulate/Fibrous Composites

Chapter Outline

8.1 Introduction 167
8.2 Level Set Method for
    Composite Materials 169
    8.2.1 Level Set
          Representation 169
    8.2.2 Enrichment
          Function 172
    8.2.3 Lumped Mass
          Matrix 173
8.3 Microstructure
    Generation 175

8.4 Material Constitutive
    Model 176
8.5 Numerical Examples 177
    8.5.1 Static Analysis 177
    8.5.2 Dynamic Analysis 181
        (1) Particulate
            Composite
            Simulation 182
        (2) Fibrous
            Composites 184
8.6 Summary 187

8.1 INTRODUCTION

Polymer matrix composites (PMCs) consist of a polymer matrix combined with a dispersed phase of inclusions intended to improve mechanical properties. Recently they have been widely used in aerospace, automotive, sports equipment, and the other engineering fields (Finegan and Gibson, 1999). Figure 8.1 shows two typical polymer matrix fibrous and particulate composites that have high specific strength and stiffness, and excellent energy absorption ability.

The macroscopic properties of composites are sensitive to the shape, size, and distribution of inclusions. While significant research efforts have been made to calculate effective elastic moduli of PMCs, dynamical behaviors in PMCs differ greatly from the static material response, particularly when the deformation wavelength is comparable to the characteristic length of the microstructure. The propagation of periodic and transient waves in PMCs is complicated due to the combination of wave scattering from material

Extended Finite Element Method. http://dx.doi.org/10.1016/B978-0-12-407717-1.00008-X
Copyright © 2014 Tsinghua University Press. Published by Elsevier Inc. All rights reserved.

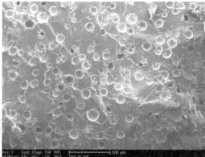

FIGURE 8.1 Polymer matrix fibrous and particulate composites.

interfaces and dissipation in the matrix (Biwa, 2001). Understanding and predicting wave attenuation in these complex materials is useful in many fields, including geophysics, oil exploration, nondestructive evaluation, and acoustic wave energy management.

Many theoretical and experimental investigations have appeared in the literature. Recently, Kim (2010) compared eight different theoretical models in a set of simple benchmark problems for predictions of effective wave speed and coherent attenuation in two-dimensional elastic composites. These benchmark results illustrate that the disagreement in attenuation can be significant, particularly at high frequencies where attention is sensitive to the composite microstructure. Biwa et al. (2002) developed a semi-analytical model to estimate the ultrasonic attenuation in fiber/particulate viscoelastic composites based on scattering and absorption losses, which expands the application of theoretical analysis in this area. Due to the mathematical complexity of these theories, several simplifying assumptions are made to make practical predictions, including neglecting the mutual interaction of scattered waves radiated from neighboring inclusions, and considering the scattering and absorption independently. The first of these assumptions limits the applicability to dilute composite concentrations. Kinra et al. (1980) reported experimental measurements of the phase velocity and attenuation of longitudinal waves in a random particulate composite in the frequency range of 0.3–5.0 MHz. Zhuang and Ravichandrana (2003) experimentally investigated the effect of interface scattering on the effective shock viscosity in a layered material. Experimental measurements of wave propagation in composites are especially difficult because, at higher frequencies, multiple-scattering noise makes identifying coherent signals challenging.

The challenges in experimental and theoretical characterization of the dynamical response of PMCs provide motivation for the development of numerical tools that can accurately predict static and dynamic properties (Segurado and Llorca, 2002; Pan et al., 2008). In the modeling of composites by conventional FEM, the explicit meshing of complex geometry such as composites is a challenging task and poor quality meshes can greatly deteriorate the critical time step in explicit time integration or result in ill-conditioning of the system, leading to spurious wave dispersion. The idea of representing weak discontinuities in a domain as enrichments to the finite element basis, rather than by forcing them to coincide with element boundaries, was first proposed by Sukumar et al (2001).

X-FEM has been successfully applied to dynamic problems, particularly for studying crack propagation, as described in the preceding chapters. In contrast to the equilibrium methods, the application of X-FEM to study the dynamic properties of composites and heterogeneous materials has not received as much attention. In order for elastodynamics analysis to be computationally tractable, the explicit integration schemes are combined with mass lumping such that no solution of a system of equations is required. However, there is not yet a sophisticated mass lumping scheme for the weak discontinuity problem. In this chapter we investigate the static and dynamic properties of particulate and fibrous composites using the X-FEM/level set method, where a lumping scheme that preserves kinetic energy is proposed. The contents are organized as follows. In section 8.2, the enrichments for weak discontinuities are reviewed and the mass lumping method used for efficient explicit time integration is described. Section 8.3 presents the method for generating inclusion distributions, where the random sequential algorithm is employed. A linear viscoelastic material model suitable for modeling epoxy and polyurea is presented in section 8.4, and numerical examples follow in section 8.5. The final section contains concluding remarks.

8.2 LEVEL SET METHOD FOR COMPOSITE MATERIALS

In finite element simulation of composite materials, inhomogeneities, such as voids and inclusions, complicate the meshing process because element faces must coincide with all material interfaces so that a weak discontinuity, i.e. a discontinuity in strain, is contained within the finite element basis. This requirement is particularly imposing for three-dimensional geometries, and limits the modeler to tetrahedral elements, which have particularly poor performance. The X-FEM can alleviate much of this burden by allowing weak discontinuities to exist independent of the mesh. This method requires a few additional steps beyond standard finite element analysis, namely the representation of material internal geometry or microstructure by level set functions and the extension of the finite element approximation with enrichment functions.

8.2.1 Level Set Representation

The level set method, which is described in Chapter 4, is a numerical technique for tracking moving surfaces. In this chapter, we only consider static interfaces. An interface $\Gamma \subset R^D$ (D being the dimension of the domain) can be formulated as the level set curve of a function f, where

$$\Gamma(\mathbf{x}) = \{\mathbf{x} \in R^D \,|\, f(\mathbf{x}) = 0\} \tag{8.1}$$

As mentioned in Chapter 4, one important example of such a function would be the signed distance function:

$$f(\mathbf{x}) = \pm \min_{\mathbf{x}_\Gamma \in \Gamma} \|\mathbf{x} - \mathbf{x}_\Gamma\| \tag{8.2}$$

where the absolute value of the function is equal to the distance between a point $\mathbf{x}$ and the nearest point on the interface Γ, and the sign of the function is chosen so that a positive value indicates that the point is outside the contour and a negative value that the point is inside the contour. If interior and exterior regions cannot be assigned to the contour, then the sign is simply arbitrarily and consistently chosen for either side of the contour. In the case of several interfaces Γ^k with $k = 1, 2, \ldots, N_{\text{int}}$, we can define a function associated with each interface k as

$$f^k(\mathbf{x}) = \pm \min_{\mathbf{x}_\Gamma^k \in \Gamma} \|\mathbf{x} - \mathbf{x}_\Gamma^k\| \tag{8.3}$$

In this chapter, spherical and cylindrical inclusions are considered. The signed distance function of a spherical inclusion is

$$f^k(\mathbf{x}) = \|\mathbf{x} - \mathbf{x}_C^k\| - r^k \tag{8.4}$$

where $\mathbf{x}_C^k$ and r^k are the center and radius of inclusion k respectively.

For cylindrical inclusions, a cylindrical surface is capped by two planar surfaces. The signed distance functions of the cylindrical and planar surfaces are

$$f_{\text{cyl}}^k(\mathbf{x}) = \sqrt{(\mathbf{x} - \mathbf{x}_C^k)^{\mathrm{T}}\left(I - \mathbf{e}^k(\mathbf{e}^k)^{\mathrm{T}}\right)(\mathbf{x} - \mathbf{x}_C^k)} - r^k \tag{8.5}$$

$$f_{\text{pla}}^k(\mathbf{x}) = \left|(\mathbf{x} - \mathbf{x}_C^k)\cdot\mathbf{e}^k\right| - h^k/2 \tag{8.6}$$

respectively, where for cylinder k, $\mathbf{e}^k$ is a unit-length axial vector of the cylinder, $\mathbf{x}_C^k$ is the center of the cylinder, r^k and h^k are the radius and height of the cylinder respectively. The signed distance function of the capped cylinder is then

$$f^k(\mathbf{x}) = \pm\min\left\{\left|f_{\text{cyl}}^k(\mathbf{x})\right|, \left|f_{\text{pla}}^k(\mathbf{x})\right|\right\} \tag{8.7}$$

where the sign of the function is negative only if both $f_{\text{cyl}}^k(\mathbf{x}) < 0$ and $f_{\text{pla}}^k(\mathbf{x}) < 0$.

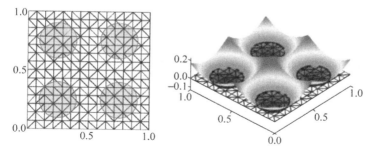

FIGURE 8.2   **Level set description of a plate with four circular inclusions (Tran et al., 2011).**

In the programming, we usually only calculate the nodal level set value, and the level set value at any position $\mathbf{x}$ inside the element can be obtained by interpolation:

$$f(\mathbf{x}) = \sum_I N_I(\mathbf{x}) f(\mathbf{x}_I) \tag{8.8}$$

where $f(x_I)$ is the level set value at node x_I. Using this formulation, it is convenient to get the derivative of level set function at point $\mathbf{x}$, which is useful in constructing the B matrix in the calculation:

$$\frac{\partial f(\mathbf{x})}{\partial \mathbf{x}} = \sum_I \frac{\partial N_I(\mathbf{x})}{\partial \mathbf{x}} f(\mathbf{x}_I) \tag{8.9}$$

Figure 8.2 shows an example of applying a level set function to describe a plate with four circular inclusions. The plate is divided into triangular elements and the mesh is independent of the material interface. It should be pointed out that, if a linear approximation is used for the zero level set approximation, the interface will result in piecewise linear geometrical elements (segments in 2D, triangles in 3D). For example, the dotted line in Figure 8.3 is the approximated intersection line of a circle and ellipse with one quadrilateral element.

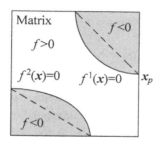

FIGURE 8.3   **Level set description when one element intersects with more than one interface.**

In Chapter 4 we talked about the subdomain integration of a discontinuous field in X-FEM, in which we need to know the position of the discontinuity. In the simulations the intersection points of the material interface and element can be obtained as follows. Let f_i and f_j denote nodal level set values at two vertices $\mathbf{x}_i$ and $\mathbf{x}_j$ of an element. An element edge is cut by the zero level set if $f_i f_j < 0$. The resulting intersection point $\mathbf{x}_p$ (see Figure 8.3) is then found by

$$\mathbf{x}_p = \mathbf{x}_i + \xi(\mathbf{x}_j - \mathbf{x}_i), \quad \xi = \frac{f_i}{f_i - f_j} \tag{8.10}$$

8.2.2 Enrichment Function

For a strongly bonded interface within a composite material, the displacement field remains continuous, but the strain can be discontinuous. Traditionally finite element models have required that these interfaces coincide with element faces so that this weak discontinuity can be represented. In the X-FEM, an enriched displacement field is developed to accommodate this weak discontinuity without requiring it to reside along element faces.

Here a modified enrichment function $\Phi(\mathbf{x})$ proposed by Moës et al. (2003) is constructed to preserve the weak discontinuity at the interface, but also vanish along all element faces that are not intersected by the interface:

$$\Phi(\mathbf{x}) = \sum_I |f(\mathbf{x}_I)| N_I(\mathbf{x}) - \left| \sum_I f(\mathbf{x}_I) N_I(\mathbf{x}) \right| \tag{8.11}$$

This construction eliminates blending elements for which only a subset of the element nodes are enriched (see Chapter 4 for details).

In the modeling of composites, one element may intersect with more than one material interface (see Figure 8.3) when the volume fraction of inclusions is high. For this situation, we need to construct an enrichment function for each interface that intersects with the same element, then the displacement field of the element is expressed as

$$\mathbf{u}^h(\mathbf{x}, t) = \sum_{I \in S} N_I(\mathbf{x}) \mathbf{u}_I(t) + \sum_k^{N_{int}} \sum_{J \in S_{interface}^k} N_J(\mathbf{x}) \Phi^k(f^k(\mathbf{x})) \mathbf{q}_J^k \tag{8.12}$$

where $S_{interface}^k$ denotes all the nodes of the element passed by the kth interface, $\mathbf{q}_J^k$ is the additional degrees of freedom corresponding to interface k, and N_{int} is the number of total interfaces. Usually, as the number of interfaces in one element increases, the accuracy of the solution will decrease. So generally we avoid this number exceeding two in the calculation.

8.2.3 Lumped Mass Matrix

By substituting the displacement approximation in Eq. (8.12) into the weak form of the momentum equation (see Chapter 4 for details), we obtain the discrete equation of dynamics:

$$\mathbb{M}\ddot{U} = f^{\text{ext}} - f^{\text{int}}, \quad U = [u \quad q] \tag{8.13}$$

where u and q are nodal unknowns and $\mathbb{M}$ is the mass matrix, f^{ext} and f^{int} are external force and internal force vectors respectively. The solution of Eq. (8.13) (including spatial integral and time integration) has been given in detail in Chapter 4. Here we focus on discussing the lumping scheme of the mass matri $\mathbb{M}$.

The mass matrix $\mathbb{M}$ for an enriched domain can be written in matrix form as

$$\mathbb{M} = \begin{bmatrix} \mathbb{M}_{11} & \mathbb{M}_{12} \\ \mathbb{M}_{12}^{T} & \mathbb{M}_{22} \end{bmatrix} \tag{8.14}$$

where $\mathbb{M}_{11}$ is the standard mass matrix corresponding to displacement unknowns, $(\mathbb{M}_{11})_{IJ} = \int_{\Omega_0} \rho_0(\mathbf{x}) N_I N_J d\Omega_0$, $\mathbb{M}_{22}$ is the enriched mass matrix corresponding to additional degrees of freedom, $(\mathbb{M}_{22})_{IJ} = \int_{\Omega_0} \rho_0(\mathbf{x}) N_I N_J \Phi^2(\mathbf{x}) d\Omega_0$, and the mixed mass matrix is expressed as $(\mathbb{M}_{12})_{IJ} = \int_{\Omega_0} \rho_0(\mathbf{x}) N_I N_J \Phi(\mathbf{x}) d\Omega_0$.

Row summation is a commonly used lumping scheme in the explicit time integration method, where the mass of element is equally distributed to the nodes. However, for heterogeneous materials such as composites, row summation is inappropriate due to the difference in material densities between the inclusions and the matrix materials. Especially for the modeling of composites with voids, the row summation lumping scheme will lead to the stable time step tending to zero (Rozycki et al., 2008). A new lumping scheme that preserves kinetic energy is proposed in this study. Here the mass matrix is diagonalized by neglecting the mixed mass terms $\mathbb{M}_{12}$ and $\mathbb{M}_{21}$, and computing the diagonalized standard and enriched mass by

$$(\mathbb{M}_{11})_{II}^{j} = \frac{1}{n_{\text{node}}} \int_{\Omega_j} \rho_0(\mathbf{x}) d\Omega_j, \quad (\mathbb{M}_{22})_{II}^{j} = \frac{1}{n_{\text{node}}} \int_{\Omega_j} \rho_0(\mathbf{x}) \Phi^2(\mathbf{x}) d\Omega_j \tag{8.15}$$

where the element is divided into subdomains Ω_j by the material interfaces, and the total mass of the subdomain is equally distributed to each node contained within the subdomain, i.e. n_{node} is the number of nodes contained within each subdomain j. The principle of the lumping mass matrix is to keep the kinematic energy of the system conserved for rigid motion mode. Next we use a 1D problem as an example to demonstrate that this lumping method satisfies this condition.

FIGURE 8.4 Description of 1D interface element with cross-sectional area A.

The one-dimensional element considered here has two nodes, as shown in Figure 8.4. The element is divided into two domains, Ω_1 and Ω_2, by the material interface. The corresponding densities are ρ_1 and ρ_2 for each domain. According to Eq. (8.12), the approximate displacement is

$$u^{\mathrm{h}}(x) = N_1(x)u_1 + N_2(x)u_2 + N_1(x)\Phi(x)q_1 + N_2(x)\Phi(x)q_2 \qquad (8.16)$$

A lumped mass matrix for this element is of the form:

$$\mathbb{M}^{\mathrm{lump}} = \begin{bmatrix} m_1 & 0 & 0 & 0 \\ 0 & m_2 & 0 & 0 \\ 0 & 0 & m_3 & 0 \\ 0 & 0 & 0 & m_4 \end{bmatrix} \qquad (8.17)$$

where m_1 and m_2 correspond to u_1 and u_2, and m_3 and m_4 correspond to q_1 and q_2. We consider a motion described by $\dot{u} = \tilde{v}\Phi(x)$. We set $\dot{u}_1$ and $\dot{u}_2$ to zero and $\dot{q}_1$ and $\dot{q}_2$ to $\tilde{v}$. The exact kinematic energy T and the kinematic energy T^{lump} expressed by the lumped mass matrix can be computed by

$$T = \frac{1}{2}\left[\int_{\Omega_1} \rho_1 \Phi^2(x)\mathrm{d}\Omega + \int_{\Omega_2} \rho_2 \Phi^2(x)\mathrm{d}\Omega\right]\tilde{v}^2 \qquad (8.18)$$

and

$$T^{\mathrm{lump}} = \frac{1}{2}[m_3 + m_4]\tilde{v}^2 \qquad (8.19)$$

According to Eq. (8.15), we can see that the kinematic energy is exactly conserved for our proposed lumping scheme. Similarly, by considering a motion described by a constant velocity $\dot{u} = \tilde{v}$, and setting $\dot{u}_1$ and $\dot{u}_2$ to $\tilde{v}$ and $\dot{q}_1$ and $\dot{q}_2$ to zero, the lumping for m_1 and m_2 can also be verified. This method fails when nodes are located within voids as it would result in nodes that contain zero mass. For this case, we can equally distribute the element mass to each node (Rozycki et al., 2008). For example, in the 1D element in Figure 8.4, the lumped mass $m_1 = m_2 = \frac{1}{2}\rho_1 l_1 A$, when $\rho_2 = 0$.

Still taking the 1D interface element of length l in Figure 8.4 as an example, we can study the stable time step calculated from the proposed mass lumping strategy. The stable time step is expressed as $t_{\mathrm{c}} = 2/\omega_{\mathrm{max}}$, where the maximum frequency ω_{max} is obtained from the eigenvalue problem $\det(\mathbf{K}^{\mathrm{e}} - \omega^2\mathbf{M}^{\mathrm{e}}) = 0$, in which $\mathbf{K}^{\mathrm{e}}$ and $\mathbf{M}^{\mathrm{e}}$ are stiffness and mass matrices of the interface element respectively. The variation of critical time steps with the position of the interface is shown in Figure 8.5 for different

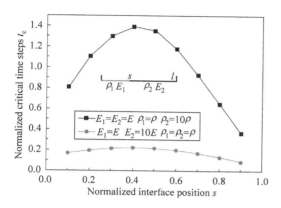

FIGURE 8.5   Critical time steps (normalized by $l\sqrt{\rho/E}$) as a function of the location of the material interface s (normalized by element length l) in a 1D interface element.

material properties. We can find that the stable time step is still acceptable even when the interface is closed to the element nodes in the proposed lumping strategy.

8.3 MICROSTRUCTURE GENERATION

The random sequential adsorption (RSA) algorithm (Feder, 1980) was used to generate particulate/fibrous composites. This algorithm prevents the occurrence of spuriously overlapping inclusions by requiring that each newly generated candidate inclusion does not overlap any previously accepted. The current volume fraction is updated on the acceptance of each new inclusion and the process continues until the desired volume fraction is met or a predefined number of attempts has been exceeded. Due to the lack of rearranging inclusions, this method can take a large number of iterations or stall when high (>50%) volume fractions are requested. Higher volume fractions of spherical particles were generated by running a molecular dynamics simulation with strongly repulsively interacting particles and selecting a time step where particles did not overlap. Due to the mesh independence of the particle geometry gained from the X-FEM, no additional steps or geometry constraints needed to be applied during the RSA or MD particle distribution process (Segurado and Llorca, 2002).

It is straightforward to generate spherical inclusion composites since the zero overlap condition between two particles only requires that the distance between sphere centers must be greater than the sum of their radii. For short fibers with small aspect ratios and low volume fractions, fibers can be modeled as straight cylinders. In three dimensions, a cylinder can be fully described by a center point x_0, radius r, height h, and two Euler angles, θ and ϕ (see Figure 8.6). The center point was chosen by a uniform random

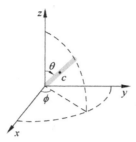

FIGURE 8.6 The description of a cylinder in 3D space.

distribution, and the Euler angles were selected with a biased distribution (Schladitz et al., 2006):

$$p(\theta, \phi) = \frac{1}{4\pi} \frac{\beta \sin \theta}{\left(1 + (\beta^2 - 1)\cos^2 \theta\right)^{3/2}} \tag{8.20}$$

where $\theta \in [0, \pi)$ and $\phi \in [0, 2\pi)$ give the altitude and longitude in spherical coordinates. The bias parameter varies from $\beta = 1$ for an isotropic distribution, or as $\beta \to 0$ fibers are biased to lie in-plane. For the intersection test of the cylinders, if the test fails, only the center point position is recalculated while the Euler angle remains unchanged.

8.4 MATERIAL CONSTITUTIVE MODEL

In the study, the glass spheres/fibers are considered as linear elastic, and the epoxy matrix is approximated as a linear viscoelastic material. A simplified version of the temperature- and pressure-dependent viscoelastic constitutive model (Amirkhizi et al., 2006) is used. In bulk deformations, this model assumes the trace of the Cauchy stress tensor σ is related to the volume ratio by

$$\text{tr}(\sigma) = 3\kappa \frac{\ln J}{J} \tag{8.21}$$

where $J = \det(\mathbf{F})$ is the Jacobian of the deformation gradient and κ is bulk modulus. A hereditary integral is employed to express the deviatoric part of stress:

$$\sigma(t) = \int_0^t \frac{T(t')}{T_{\text{ref}}} 2\mu(\tau - \tau') \dot{e} dt' \tag{8.22}$$

where T_{ref} is the reference (room) temperature, $\dot{e}$ is the deviatoric part of the deformation rate tensor, and τ is a reduced time, related to the actual time, t, by the following equation:

$$\tau = \int_0^t \frac{dt}{a_T(t)}, \quad a(T) = 10^{\frac{A\left(T - T_{ref}\right)}{B + \left(T - T_{ref}\right)}} \tag{8.23}$$

The relaxation modulus $G(t)$ is defined by a Prony series expansion:

$$G(t) = \mu_0 \left(1 - \sum_{i=1}^n \left(1 - p_i e^{-t/q_i} \right) \right) \tag{8.24}$$

where μ_0 is instantaneous shear modulus, p_i and q_i are material constants. The local temperature is assumed adiabatic and is updated by

$$\frac{\partial T}{\partial t} = \frac{1}{C_v} \frac{\partial W_d}{\partial t} \tag{8.25}$$

where C_v is the heat capacity and W_d is the dissipated work.

8.5 NUMERICAL EXAMPLES

In this section, the X-FEM/level set method is applied to simulate the static and dynamic behaviors of composites consisting of spherical/cylindrical inclusions distributed in an epoxy matrix.

8.5.1 Static Analysis

First the reliability of the algorithm is verified by a simple uniaxial tension simulation for a cubic material with a sphere inclusion at the center. The side length is 1.0 mm and the diameter of the inclusion is 0.2 mm, as shown in Figure 8.7. The sphere inclusion is linear elastic: Young's modulus is 80 GPa, Poisson's ratio is 0.3, the epoxy matrix is viscoelastic material, and the material constants are given in Table 8.1.

This problem is simulated by conventional FEM and X-FEM (Figure 8.8). Figure 8.8(a) shows the mesh of the commercial FEM software ABAQUS, in which the element boundaries coincide with the material interface; Figure 8.8(b) is the regular hexahedron mesh used in X-FEM.

The homogenized (or macro) stress during the tension is expressed as

$$\langle \sigma \rangle = \frac{1}{\Omega} \int_\Omega \sigma(x) d\Omega \tag{8.26}$$

where Ω is the volume of the material domain. The homogenized stress–strain curves obtained by X-FEM compare well with those of the FEM, as shown in

FIGURE 8.7 Comparison of homogenized tension stress–strain curves.

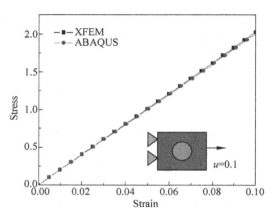

TABLE 8.1 Values of the Constitutive Parameters in the Epoxy Matrix Model

T_{ref} (K)	A	B (K)	C_v (J/mm³/K)	n	K (GPa)	G_0 (GPa)	ρ (g/cm³)
273	-10	107.54	1.977×10^{-3}	4	5.205	1.4818	1.18

p_1	p_2	p_3	p_4	q_1	q_2	q_3	q_4
0.0738	0.1470	0.3134	0.3786	463.4	0.06407	0.0001163	7.321×10^{-7}

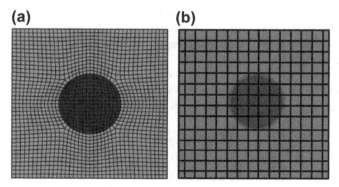

FIGURE 8.8   Meshes used in conventional FEM (a) and X-FEM (b).

FIGURE 8.9   Contours of tension strain (top: FEM; bottom: X-FEM).

Figure 8.7. Furthermore, the contours of the tension strain field of both methods are shown in Figure 8.9. For ease of observation, the cut views are also given in Figure 8.9, from which we can find the discontinuity of the strain field at the interface of inclusion and matrix.

Under cyclic loading, the stress—strain curve of viscoelastic material shows a hysteresis because of the phase lag between stain and stress. By applying sinusoidal loading, the homogenized hysteresis stress—strain curve can be obtained and plotted in Figure 8.10, from which we can find that adding hard sphere inclusion can increase the energy absorption ability of pure polymer matrix.

The next example is to simulate the uniaxial tension for composites with unidirectional fibers. The three sets of samples shown in Figure 8.11 are considered in the simulations: fiber direction aligned with loading direction; fiber direction perpendicular to loading direction; no fibers. The material constants of the fiber and matrix are the same as in the last example. The macro stress—strain curves for these three models are shown in Figure 8.12, from which we can see that the effective modulus along the loading direction

FIGURE 8.10 Comparisons for homogenized hysteresis stress–strain curves of polymer with and without inclusion.

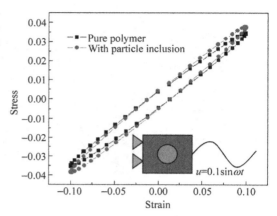

FIGURE 8.11 Three groups of samples. Left to right: fiber direction aligned with loading direction, perpendicular to loading direction, and pure polymer.

FIGURE 8.12 Homogenized stress–strain curves of three sets of samples.

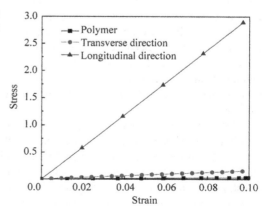

FIGURE 8.13   Contour of tension strain when the fibers are perpendicular to the loading direction.

is greatly improved for sample 1. The contour of tension strain for sample 2 is given in Figure 8.13.

8.5.2 Dynamic Analysis

In this section, the X-FEM/level set method is applied to simulate ultrasonic wave propagation in a composite consisting of spherical/cylindrical inclusions distributed in an epoxy matrix and quantitatively study the wave attenuation in viscoelastic composites (Liu et al., 2013).

The wave attenuation can be quantitatively computed by the longitudinal attenuation coefficient α, which is defined as

$$\alpha_L = \ln(S_1/S_2)/(x_1 - x_2) \text{ nepers/mm} \tag{8.27}$$

where S_1 and S_2 are the surface averaged stresses along the wave propagation direction at positions x_1 and x_2 respectively.

The viscoelastic material model in section 8.4 is used for epoxy matrix. The viscosity of the bulk modulus is neglected and only the viscosity of the shear modulus is considered in the calculation. The constants in Prony series (see section 8.4) are obtained by fitting with experiment data. The Prony series terms are calibrated using frequency-dependent test data. The complex moduli can be obtained from the wave velocity and attenuation as a function of the angular frequency ω. For an isotropic viscoelastic material, the complex Lamé parameter λ and $\mu = G_s + iG_l$ can be identified via the following well-known relations:

$$\lambda + 2\mu = \rho C_L^2 \left(1 + i\frac{C_L \alpha_L}{\omega}\right)^{-2} \tag{8.28}$$

where ρ is density, C_L and α_L are the speed and attenuation coefficient of a longitudinal wave respectively, which are experimentally measured at different frequencies. On the other hand, the complex shear moduli μ can also be obtained by converting the Prony series terms from time domain to the frequency domain, and written as follows:

$$G_s(\omega) = G_0 \left[1 - \sum_{i=1}^{n} p_i \right] + G_0 \sum_{i=1}^{n} \frac{p_i q_i^2 \omega^2}{1 + q_i^2 \omega^2}$$

$$G_l(\omega) = G_0 \sum_{i=1}^{n} \frac{p_i q_i^2 \omega}{1 + q_i^2 \omega^2} \tag{8.29}$$

By fitting Eq. (8.29) with Eq. (8.28), the Prony series parameters p_i and q_i are determined.

(1) Particulate Composite Simulation

Kinra et al. (1980) experimentally measured the longitudinal attenuation coefficient of a glass-particle-reinforced epoxy composite using a tone-burst ultrasonic wave in the range of 0.4–3.0 MHz. The particle radius r in the tests was 150 μm. The experimental data are for the case when the particle volume fractions c is 8.6%. In this section we simulate this experiment using X-FEM modeling and compare with the experimental data.

The geometry of the specimen used in the simulations is a $1 \times 1 \times 2$ mm^3 cuboid, as shown in Figure 8.14(a). The spherical glass inclusions are generated by the RSA process described in section 8.3. Assuming that the particles are arranged in a simple cubic structure, the interparticle distance d can be approximately estimated by $d = r(4\pi/3c)^{1/3}$. For the volume fraction above, d is 547 μm. The boundary conditions are illustrated in Figure 8.15: a uniform sinusoidal pressure with $p_0 = 10$ MPa, frequency $f = 1–4$ MHz is applied at the surface $x = 2.0$ mm along the X direction, the surface $x = 0.0$ is free, and all the other surfaces are fixed in the Y and Z directions to simulate a one-dimensional impact. The problem is solved with structure meshes consisting of cubic elements with edges of length 25 μm. The simulation time is

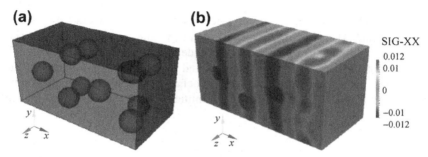

FIGURE 8.14 (a) Particulate composite with volume fraction 0.086. (b) Contour of σ_{xx} at $z = 0.95$.

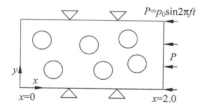

$P = p_0 \sin 2\pi f t$

FIGURE 8.15 Boundary conditions for the spherical inclusion problem. The surface $x = 0$ is defined as a traction-free surface, the surface $x = 2$ mm is loaded with a pressure P, and all other surfaces are prescribed with zero normal displacement and zero shear tractions.

1×10^{-3} ms, time step $\delta t = 1e^{-6}$ ms. The elastic material parameters for glass inclusion are: Young's modulus $= 64.89$ GPa, Poisson's ratio $= 0.249$, and density $= 2.47$ g/cm^3; the viscoelastic material model in section 8.4 is used for epoxy matrix. Since the temperature fluctuation in the experiments is small, the temperature effect in the viscoelastic material model is disabled in our simulations. The constants in Prony series are given in Table 8.1, which are obtained by fitting with experiments. The viscosity of the bulk modulus is neglected.

The time histories of surface averaged stress σ_{xx} at the surfaces $x = 2.0$ and $x = 0.5$ for composites with volume fraction 8.6% are shown in Figure 8.16. We find that the peaks of the stress σ_{xx} at $x = 0.5$ decrease compared with those at $x = 2.0$ due to scattering by the inclusions and absorption by the matrix, and this decrease grows strongly with increasing frequency. The calculated results are plotted in Figure 8.17 together with published theoretical predictions (Biwa, 2001) and experimental measurements (Kinra et al.,

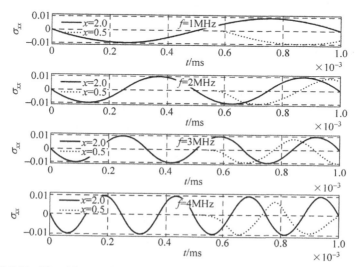

FIGURE 8.16   Time history of surface averaged σ_{xx} at $x = 2.0$ and $x = 0.5$ for different frequencies.

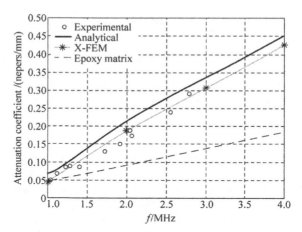

FIGURE 8.17   Attenuation in the particulate composite vs. frequency.

1980). The theoretical analysis (Biwa, 2001) overestimates the attenuation, especially for the high volume fraction composites. This is because in the analytical method, multi-scattering is neglected, and becomes a worse approximation as the wavelength approaches the interparticle distance. As shown in Figure 8.17, the attenuation coefficient at lower frequencies is far less sensitive to the volume fraction, and close to that of the matrix material. After 2 MHz, the different volume fraction composites diverge significantly. The attenuation for the matrix material alone is also shown, which illustrates the dramatic increase in wave attenuation properties even at small volume fill ratios. The contour of σ_{xx} (cut at surface $z = 0.95$ mm) is shown in Figure 8.14(b).

In the experiment, the longitudinal wave speed in pure epoxy hardly changes with higher frequencies. However, after adding some inclusions into matrix, wave dispersion (wave speed changes with frequency) clearly occurs. The longitudinal group velocity C normalized by wave speed in the matrix C_m at different frequencies is plotted in Figure 8.18; the wave speed C increases with frequency, which agrees well with experimental observations. The frequency-dependent complex moduli of the composite can be obtained from the group velocity C_L and attenuation coefficient α_L according to Eq. (8.28).

(2) Fibrous Composites

Three fibrous composites with the same volume fraction (11%) were generated using the RSA process: in material (a), the orientations of the fibers are distributed isotropically; in material (b), the fibers are distributed along the X direction; in material (c), the fibers are along the Y direction, as shown in Figure 8.19. The center positions of all three sets are generated randomly. The

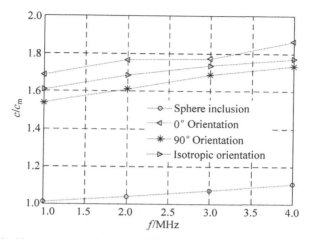

FIGURE 8.18   Normalized velocity in the different composites vs. frequency.

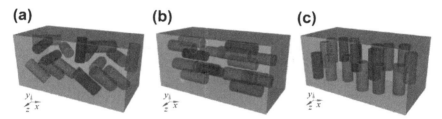

FIGURE 8.19   Three fibrous composites with the same volume fraction (11%) and different fiber orientations.

radius of fiber is 100 μm, the length is 500 μm. All the other simulation parameters, like the mesh, boundary conditions, and material model, are the same as those in particulate composite modeling.

The time histories of surface averaged stresses σ_{xx} at $x = 2.0$ and $x = 0.5$ for fibrous composite material (a) are given in Figure 8.20. The wave travels faster in the fiber, so that spreads the energy of the wave in the time domain. σ_{xx} is redistributed along the section, which decreases the surface averaged stress. A comparison of attenuation coefficients for these three materials is shown in Figure 8.21. The sample of zero degree orientation gives the best attenuation result. The contours of σ_{xx} for the sample of zero degree orientation are shown in Figure 8.22. The stress in the fibers is higher than that in the matrix, which can be attributed to the mutual interaction of the scattered waves radiated from neighboring fibers. The temperature distribution in the fibers during wave propagation is also given in Figure 8.22.

The longitudinal group velocity normalized by wave speed in the matrix C_m at different frequencies is plotted in Figure 8.18. The longitudinal modulus

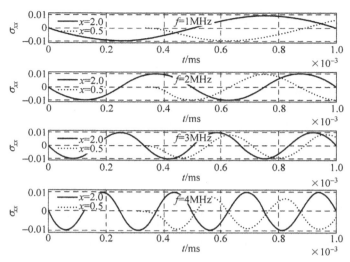

FIGURE 8.20 Time history of surface averaged σ_{xx} at $x = 2.0$ and $x = 0.5$ for different frequencies (for material (a))

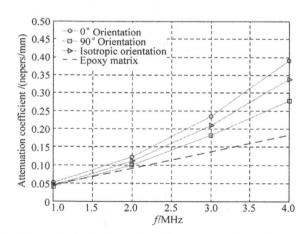

FIGURE 8.21 Attenuation in the three sets of fibrous composites vs. frequency.

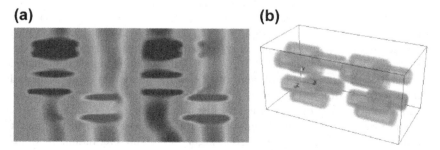

FIGURE 8.22 Contours of stress and temperature in material (b).

of the zero degree aligned fibers is the largest, as demonstrated by the highest group velocity. The fibers act to increase wave dispersion because they locally increase the wave velocity and cause the wave energy to become more spread out in the domain.

8.6 SUMMARY

The extended finite element method (X-FEM) has been demonstrated to successfully predict static and dynamic properties of polymer matrix particulate/fibrous composites. In comparison with analytical predictions and experimental measurements of the wave attenuation coefficients in epoxy/glass particulate composites the X-FEM numerical model shows superior agreement with the experimental data in the range of 1–4 MHz, as the wavelength approaches the mean particle spacing. The primary advantage of this approach over analytical methods is the ability to predict the effects of inclusions that are not dilute, have complicated geometry, or that are not randomly dispersed or oriented in the matrix.

The main advantage of the X-FEM in modeling composite behavior is the ease with which high-quality computational meshes can be generated for arbitrary inclusion geometries and distributions. While we have generated arbitrary random microstructures in this study using the RSA algorithm, it would be straightforward to use reconstructive techniques and X-ray tomography to model specific polymeric unit cells, or use these measurements to better statistically describe the inclusion distribution in a composite.

For fiber composites, the X-FEM approach for computing the WAC shows that maximum attenuation can be achieved by aligning fibers with the direction of wave propagation for longitudinal waves. In this configuration, the WAC increased by more than a factor of two at 4 MHz, the highest frequency considered. These results illustrate the potential of the X-FEM approach in the modeling of composites with complex geometries.

Chapter 9

X-FEM Simulation of Two-Phase Flows

Chapter Outline

9.1 Governing Equations and Interfacial Conditions   189
9.2 Interfacial Description of Two-Phase Flows   192
9.3 X-FEM and Unknown Parameters Discretization   194
9.4 Discretization of Governing Equations   200
9.5 Numerical Integral Method   205
9.6 Examples and Analyses   207
9.7 Summary   211

Two-phase flows are commonplace in daily life. Examples can be found in marine engineering, wave dynamics, chemical engineering, the food processing industry, etc. In order to allow for the complexity of real-life two-phase flows, elaborate treatments are desirable for numerical modeling and simulation. To this end, the numerical simulation of transient immiscible and incompressible two-phase flows is proposed in this chapter, demonstratings how to deal with multi-phase flow problems by applying the X-FEM. Flow simulation in the two-dimension spatial domain is only taken into account here. However, the potential exists for the method to be extended to three-dimension flow simulation.

9.1 GOVERNING EQUATIONS AND INTERFACIAL CONDITIONS

The domains occupied by two distinct kinds of immiscible fluids, namely fluids 1 and 2, are denoted by Ω_1 and Ω_2 respectively. Their union is defined as $\Omega = \Omega_1 \cup \Omega_2$ and is shown in Figure 9.1.

Because the flow is incompressible, the ratio of fluid volume change is zero, while the volume change ratio of continuous material can be characterized by the velocity divergence. Thus, the mass conservation equation in

Extended Finite Element Method. http://dx.doi.org/10.1016/B978-0-12-407717-1.00009-1

FIGURE 9.1 Solution domain occupied by two kinds of immiscible fluids.

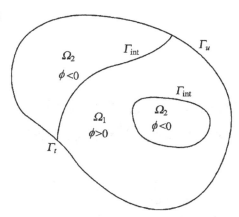

Euler form can be simplified to be the condition of velocity divergence equivalent to zero, such that

$$\nabla \cdot \boldsymbol{u} = 0 \tag{9.1}$$

where $\boldsymbol{u}$ is the velocity vector in the fluid field. In each computational domain, the flow from the two kinds of fluid is satisfied by the same form of momentum equation, which is

$$\rho \left(\frac{\partial \boldsymbol{u}}{\partial t} + \boldsymbol{u} \cdot \nabla \boldsymbol{u} \right) = \rho \boldsymbol{f} + \nabla \cdot \boldsymbol{\sigma} \tag{9.2}$$

where ρ is fluid density and f is the volumetric force in a unit mass of fluid, for example gravity. σ is the total stress of the fluid field. For incompressible flow, the fluid stress can be written as

$$\boldsymbol{\sigma} = \boldsymbol{\tau} - p\boldsymbol{I} \tag{9.3}$$

in which τ is viscosity stress and p is pressure (also called intensity of pressure). The fluid considered here is incompressible Newtonian flow. The relation between viscosity stress τ and deformation ratio tensor $\boldsymbol{D}$ $(\boldsymbol{D} = \frac{1}{2}(\nabla \boldsymbol{u} + \boldsymbol{u}\nabla))$ is linear constitutive, namely

$$\boldsymbol{\tau} = 2\mu \boldsymbol{D} = \mu(\nabla \boldsymbol{u} + \boldsymbol{u}\nabla) \tag{9.4}$$

where μ is the dynamic viscosity coefficient. Isothermal flow only is considered, such that μ is a constant.

If the action of surface tension f_{ST} at the two-phase interface Γ_{int} is considered, then the surface tension is given by

$$\boldsymbol{f}_{ST} = \gamma \kappa \delta(\boldsymbol{x} - \bar{\boldsymbol{x}})\boldsymbol{n} \quad \forall \boldsymbol{x} \in \Omega \tag{9.5}$$

where γ is a coefficient of surface tension in the normal direction, which is assumed to be constant here; κ is the average curvature of the two-phase

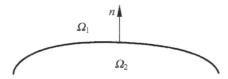

FIGURE 9.2   Normal direction.

interface; $\delta(x - \bar{x})$ is the Dirac δ function defined in the two-phase interface Γ_{int}, in which $\bar{x}$ is the closest point on the interface to the point x; n denotes the interfacial unity normal vector from Ω_2 to Ω_1 in the positive orientation, as shown in Figure 9.2.

Across the interface Γ_{int} of two-phase flows, the kinematic and kinetic continuity conditions must be satisfied by the fluid field. The force balance is firstly considered at the interface. Through the definition of surface tension, the relation between fluid stress and surface tension at the interface can be obtained as follows:

$$[\![p\boldsymbol{I} - \boldsymbol{\tau}]\!] \cdot \boldsymbol{n} = \gamma\kappa\boldsymbol{n} \qquad \forall x \in \gamma_{int} \tag{9.6}$$

where $[\![*]\!] = [\![*]\!]|_{\Omega_1} - [\![*]\!]|_{\Omega_2}$. It can be seen from the equilibrium condition of interfacial forces that the total stress generated by two-phase flow is balanced with the surface tension determined by the surface character and curvature along the normal direction of the interface. In the projection of viscosity stress to the interface, the tangential component is continuous, which physically corresponds to the kinematic continuity condition of the interface.

In addition, there is no sliding along the tangential direction of the interface since the two viscous fluids are immiscible to each other. There is also no mass exchange along the normal direction. Thus, it follows the no sliding kinematic continuity condition as the interface, which is

$$[\![\boldsymbol{u}]\!] = 0 \qquad \forall x \in \gamma_{int} \tag{9.7}$$

In order to find the solution, the fluid field must satisfy the no sliding boundary and force equilibrium conditions except for the two-phase interfacial condition; that is,

$$\boldsymbol{u} = \bar{\boldsymbol{u}} \qquad \forall x \in \gamma_u \tag{9.8}$$

$$\boldsymbol{\tau} \cdot \boldsymbol{n} - p\boldsymbol{n} = \bar{\boldsymbol{t}} \qquad \forall x \in \gamma_t \tag{9.9}$$

where $\bar{u}$ and $\bar{t}$ are the specified velocity at velocity boundary Γ_u and the specified stress at stress boundary Γ_t respectively, as shown in Figure 9.1.

Now, the governing Navier–Stokes (N-S) equations of the fluid field are established, as well as the interface, boundary, and initial conditions, which are summarized below:

$$\rho\left(\frac{\partial u}{\partial t} + u \cdot \nabla u\right) = \rho f + \gamma \kappa \delta(x - \bar{x})n + \nabla \cdot (\mu(\nabla u + u\nabla)) - \nabla p \qquad (9.10)$$

and mass conservation condition,

$$\nabla \cdot u = 0$$

the no sliding boundary condition,

$$u = \bar{u} \quad \forall x \in \gamma_u$$

the force equilibrium condition,

$$\tau \cdot n - pn = \bar{t} \quad \forall x \in \gamma_t$$

the no sliding interfacial condition,

$$[\![u]\!] = 0 \quad \forall x \in \gamma_{\text{int}}$$

as well as the interfacial force equilibrium condition,

$$[\![pI - \tau]\!] \cdot n = \gamma \kappa n \quad \forall x \in \gamma_{\text{int}}$$

9.2 INTERFACIAL DESCRIPTION OF TWO-PHASE FLOWS

For transient two-phase flows, the position of the interface varies with time. In order to simulate two-phase flows accurately, the fluid field behavior near the interface must be captured precisely. Thus, the correct position of the interface is found. If the interface position is found at each time, two methods can be adopted: the interface tracking method and the interface capturing method. Interface tracking is an explicit method. The Lagrange mesh is first constructed to match the interface, and the mesh moves to follow the interface variation. Then, according to the physical features and interface movement characteristics of the problem, the interfacial position is found explicitly. However, interface capture is an implicit method. The function on the background mesh is first defined and function evolution is carried out. Then, by dealing with the function value, for example, interpolation, etc., the interfacial position is determined.

A fixed Euler mesh is used as the background mesh for most fluid field simulations. The implicit function constructed in the Euler mesh is much simpler than the mesh construction following the movement of the interface. Therefore, the Euler mesh and the level set function are used here to describe the interface. According to the interfacial position, the function values of level set may have many definitions. However, one of the most straightforward methods is to define the function value as a direction distance from the spatial point to the interface. What is called the direction distance is a signed distance function, which is not only dependent on the absolute value of the shortest distance from the spatial point to the interface, but also depends on the relative location from this point to the interface.

A level set function $\phi(x)$, which corresponds to the arbitrary spatial point x, is defined in the entire domain Ω. The spatial point located at the interface is $\overline{x}$. The definition method of direction distance continues to be used as given in Figure 9.2; that is,

$$|\phi(x)| = \min_{\overline{x} \in \gamma_{\text{int}}} \|x - \overline{x}\| \tag{9.11}$$

$$\text{sign}(\phi(x)) = \begin{cases} \dfrac{x - \overline{x}_\phi}{\|x - \overline{x}_\phi\|} \cdot n_\phi, & x \neq \overline{x}_\phi \\ 0, & x = \overline{x}_\phi \end{cases} \tag{9.12}$$

where $\overline{x}_\phi$ is a point located at the interface Γ_{int}, which makes $\phi(x)$ satisfy the minimum condition of Eq. (9.11). n_ϕ is the unit normal direction at the interface of $\overline{x}_\phi$. In other words, the level set function can be reformulated as

$$\phi(x) = (x - \overline{x}_\phi) \cdot n_\phi, \quad \forall x \in \Omega$$

The level set definition is illustrated in detail in Figure 9.3. It can be seen that the level set value is zero at the interface, such that

$$\phi(x) = 0 \quad \forall x \in \gamma_{\text{int}} \tag{9.13}$$

Based on the properties of the signed level set, the fluid properties can be defined as a function of spatial coordinates x as

$$\rho(x) = \rho_2 + (\rho_1 - \rho_2)H(\phi(x)) \quad \forall x \in \Omega \tag{9.14}$$

$$\mu(x) = \mu_2 + (\mu_1 - \mu_2)H(\phi(x)) \quad \forall x \in \Omega \tag{9.15}$$

where $H(*)$ is the Heaviside function:

$$H(x) = \begin{cases} 1, & x \geq 0 \\ 0, & x < 0 \end{cases} \tag{9.16}$$

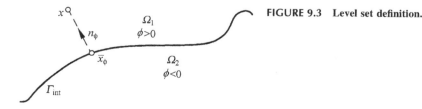

FIGURE 9.3   Level set definition.

In a transient analysis, particles on the interface initially are hypothesized to stay on it at all times in immiscible two-phase flows. As a result, following Eq. (9.13), the material derivative of the level set ϕ with respect to time t remains zero persistently, i.e.

$$\frac{D\phi}{Dt} = 0$$

That is,

$$\frac{\partial \phi}{\partial t} + \boldsymbol{u} \cdot \nabla \phi = 0 \quad \forall x \in \gamma_{\text{int}} \tag{9.17}$$

where $D(*)/Dt$ indicates the material derivative of physical quantity, which consists of a partial time derivative $\partial(*)/\partial t$ and a convective velocity $\boldsymbol{u} \cdot \nabla(*)$. Therefore, the solution of Eq. (9.17) can be achieved when the proper initial conditions and boundary conditions are provided to the level set. As the result, the interfacial position can be captured through the value of level set and the values of flow parameters in an arbitrary spatial point can be determined. Consequently, the numerical computation can be carried out.

9.3 X-FEM AND UNKNOWN PARAMETERS DISCRETIZATION

When solving N-S equations, several issues should be carefully addressed. It is necessary to consider the discontinuity situation of primary variables and unknown quantities existing at the interface. It is necessary to suppress the instability resulting from the convection term and the incompressibility in N-S momentum equations, as well as level set evolution equations. It is necessary to establish the integral scheme fully considering the effect on the interface. It is necessary to apply linear elements to characterize high-order derivative terms. In what follows, we will endeavor to develop methods to cope with these issues.

In the following discussions, if a function itself is discontinuous at the interface, it is called a strong discontinuity; if a function itself is continuous, but its derivative (no matter what order of derivative) is discontinuous at the interface, it is called a weak discontinuity, which is the same concept described in Chapter 1. The velocity $\boldsymbol{u}$ and pressure p are considered to be the unknown variables of fluid field. Through the interfacial condition (9.7), we understand that the velocity $\boldsymbol{u}$ is continuous at the interface of two-phase flows, while the interfacial condition (9.6) indicates that the components of viscosity stress τ are continuous along the interface tangential directions. However, the dynamic viscosity coefficients μ of these two kinds of fluid are usually not equivalent. Therefore, it can be understood from constitutive relations that the velocity gradient ∇u or $u\nabla$ is discontinuous at the interface. Since the viscosity stress τ does not counteract the surface tension along the normal direction of interface, the surface pressure p is generally discontinuous at the interface. In other words, the velocity u exhibits weak discontinuity and pressure p exhibits strong discontinuity at the interface of two-phase flows.

For the treatment of discontinuity, there are usually two methods. One is to treat the entire solution level without modifying any interpolation format. Smolianski (2005) used the same shape function, without any modification, for velocity and pressure to simulate incompressible two-phase flows. On the basis of the interfacial position of two-phase flows, he modified the numerical integral scheme, such that the mass quantity of flow and viscosity flux could be calculated precisely on the background elements of the located interface. Thus, the interfacial condition is satisfied in the integral to guarantee reasonable results obtained by the numerical simulation. However, because there is a limitation of interpolation orders, the numerical results cannot completely express the discontinuity of the physical quantity at the interface. The numerical results of velocity and pressure fields have a transition layer near the interface, in which their results are inaccurate. The size of this layer is affected by the factors of interpolation order, mesh density, solution scheme, etc. In order to increase the precision of numerical solution, it is necessary to reduce the size of this layer, which can be realized by refining the mesh, increasing the interpolation orders of mesh near the interface, or using a self-adaption mesh. However, the refining mesh encounters a pressure stability problem in isolated computational methods, so it should be used carefully. In addition, since there is a high gradient near the interface, where the critical situation is discontinuity, it cannot be differentiated by a low-order interpolation, like linear interpolation. Thus, the oscillation may occur near the interface. In this situation, it is necessary to have the aid of shock wave seizing technology to increase the viscosity near the interface. Therefore, the high gradient zone is limited to a small locality area in order to increase the numerical stability. However, regardless of increasing mesh density and added viscosity, this is only to restrict the transition layer to a limited zone. It does not eliminate this layer. The treated interface is still vague and inaccurate to a certain extent.

Another method is to modify the locality interpolation function, which is represented by X-FEM. At an early stage of X-FEM development, the treatment method was mainly proposed for the weak discontinuity in velocity during simulation of two-phase flows. Based on the temperature field treatment method in the phase transfer problem (Chessa et al., 2002), and according to the nodal values of the level set, an enriched shape function to express the weak discontinuity in velocity was introduced by Chessa and Belytschko (2003a). The original linear FEM domain has been extended.

Since the enriched degree of freedom corresponds to the enriched shape function, the discrete velocity field of the FEM is given by

$$u^h(x,t) = \sum_{I \in N_h} N_I(x)u_I(t) + \sum_{J \in N_h^{enr}} N_J(x)(|\phi(x)| - |\phi(x_J)|)a_J(t) \quad (9.18)$$

where $u^h(x,t)$ is the discretized fluid velocity field, N_I is the shape function of node I, N_h is a set consisting of all nodes, and N_h^{enr} is a set of nodes required by adding the enriched DOF, which indicates the nodal set on all elements crossed by the interface. In the following discussion, they are called interfacial background elements. $u_I(t)$ and $a_J(t)$ are the conventional and enriched velocity DOFs respectively. It can be verified that the value of nodal velocity x_K is

$$u(x_K) = u_K \quad (9.19)$$

This is the discretized velocity field, which is satisfied by the Kronecker δ character. However, the value of the level set is zero at the interface, so the velocity at the point of interface $\bar{x}$ is given by

$$u^h(\bar{x},t) = \sum_{I \in N_h} N_I(\bar{x})u_I(t) - \sum_{J \in N_h^{enr}} N_J(\bar{x})|\phi(x_J)|a_J(t) \quad (9.20)$$

such that

$$[\![u^h(\bar{x},t)]\!] = 0 \quad \forall \bar{x} \in \gamma_{int} \quad (9.21)$$

Since

$$\nabla u^h = \sum_{I \in N_h} \nabla N_I(x)u_I + \sum_{J \in N_h^{enr}} [\nabla N_J(x)(|\phi(x)| - |\phi_J|)a_J$$
$$+ \nabla\phi(x)N_J(x)\text{sign}(\phi(x))] \quad (9.22)$$

and

$$n_{int} = \frac{\nabla\phi}{\|\nabla\phi\|} \approx \nabla\phi \quad \forall \bar{x} \in \gamma_{int} \quad (9.23)$$

thus

$$[\![\nabla u^h(\bar{x})]\!] \cdot n_{int} = 2 \sum_{J \in N_h^{int}} N_J(\bar{x})a_J \quad \forall \bar{x} \in \gamma_{int} \quad (9.24)$$

It can be seen that the interfacial velocity continuity condition (9.7) is satisfied by the discrete velocity field expressed in Eq. (9.18), while the velocity gradient is discontinuous at the interface. The enriched shape function of a linear triangular element is a second-order function of the special coordinate. That is,

$$\psi_J(x) = |\phi(x)| - |\phi(x_J)| \quad (9.25)$$

However, there is one layer of special elements on both sides of the interfacial background elements, which are called blending elements in the following discussion. Only a part of the nodes in each blending element has an enriched DOF. Thus, in these elements, the partition of unity of the shape function is not satisfied as

$$\sum_{J \in N_h^{enr}} N_J(x) \neq 1$$

This characteristic affects the convergence of the X-FEM solution in these elements. Consequently, the accuracy is reduced and a negative effect is introduced into the solution. For this reason, a weighting modification term was added by Fries (2008) in the enriched shape function. Meanwhile, N_h^{enr} is extended to all nodes of the blending elements. The discretized velocity field of these nodes becomes

$$u^h(x, t) = \sum_{I \in N_h} N_I(x)u_I(t) + \sum_{J \in N_h^{enr}} N_J(x)\psi_J(x)A(x)a_J(t) \tag{9.26}$$

where

$$A(x) = \sum_{K \in N_h^{int}} N_K(x) \tag{9.27}$$

in which N_h^{int} indicates the nodal set on all elements crossed by the interface, which is the original N_h^{enr}. It can be verified that

$$\sum_{J \in N_h^{int}} N_J(x) = 1 \tag{9.28}$$

in the blending element. Consequently, the optimum convergence of the finite element solution is guaranteed.

In order to avoid dealing with blending elements, Moës et al. (2003) proposed an enriched shape function with a ridge form. This kind of enriched shape function only exists in the interfacial background element. The form is

$$n_J^{enr} = n_J(x) \left(\sum_I n_I(x)|\phi_I(x)| - \left| \sum_I n_I(x)\phi_I \right| \right) \tag{9.29}$$

It is easily verified that the velocity field obtained from this enriched shape function interpolation has the same Kronecker δ characteristic, as well as satisfying interfacial velocity continuous condition (9.7) and interfacial velocity gradient discontinuity. This type of interpolation method has been applied to the simulation of quasi-static multi-phase flows (Zlotnik and Diez, 2009).

In addition, the bubble function was used by Minev et al. (2003) to construct the enriched shape function, and then the enriched DOF is assembled at the element level. The second-order tetrahedral element and interface tracking method are used to simulate 3D incompressible two-phase flows. Also, a special pressure field was introduced by Coppola-Owen and Codina (2005), which is patch linear in the element and discontinuous between the elements, to simulate the weak discontinuous pressure problem.

For the strong discontinuity in pressure, the Heaviside function is adopted to construct the enriched shape function. The computational steps are consistent with the method used in the simulation of crack surface displacement discontinuity (see Chapter 4). The corresponding discretized pressure field $p^h(x,t)$ is given by

$$p^h(\boldsymbol{x}, t) = \sum_{I \in N_h} N_I(\boldsymbol{x})p_I(t) + \sum_{J \in N_h^{\mathrm{enr}}} N_J(\boldsymbol{x})(H(\phi(\boldsymbol{x})) - H(\phi(\boldsymbol{x}_J)))b_J(t) \quad (9.30)$$

where N_h^{enr} is a set of all the nodes of the interfacial background elements, which is the same definition as Eq. (9.18). $p_I(t)$ and $b_J(t)$ are the DOFs of conventional and enriched pressure respectively. It can be seen that the pressure is discontinuous at the interface, which is expressed as

$$[\![p^h(\overline{x}, t)]\!] = \sum_{J \in N_h^{\mathrm{enr}}} N_J(\overline{x})b_J(t) \quad \forall \overline{x} \in \gamma_{\mathrm{int}} \quad (9.31)$$

By means of Eq. (9.30), the numerical simulation was carried out by Reusken (2008) to solve the Laplace−Young equation, which is used to describe the suspension bubble. The relationship between the differential pressure of internal and external bubble and surface tension is reproduced precisely. Mier-Torrecilla et al. (2010) have also used Eq. (9.30) and self-adaption meshes to simulate transient immiscible and incompressible two-phase flows. It can be seen from these works that although the weak discontinuity can be treated using self-adaption meshes, it is necessary to make use of an enriched shape function dealing with an interpolation scheme to properly capture the strong discontinuity.

The enriched shape functions discussed above are capable of describing the weak discontinuity in velocity and strong discontinuity in pressure. Therefore, we can utilize them to simulate incompressible two-phase flows.

Through numerical experiments, however, we have observed that correct solution of the fluid field cannot be achieved if the problem is discretized using linear triangular finite elements and the existing weak discontinuity velocity field is modeled using Eq. (9.26). It has been recognized that the reason for this is because the incomplete high-order terms of the spatial coordinate are introduced into the enriched shape function. These high-order terms are represented only partially with respect to terms of the same order; for example, there is only an x^3 term in third-order polynomials and the x^2y, xy^2 and y^3 terms are lacking. Firstly, these incomplete interpolation functions do not reflect the rigid body motion of unknown variables and cannot express the linear behavior, while the degrees of freedom relating to these shape functions are not involved in the computation. Otherwise, errors are introduced, which results in the FEM solution not converging to the correct answer. Consequently, interfacial motion cannot be described precisely. Secondly, during the interfacial motion process, the discontinuous interface may be very close to some node or some edge of the element, which results in a great disparity between the mass corresponding to the extended DOF and the mass corresponding to the other DOF in total mass matrix. The error from this disparity is continuously accumulated, and finally results in solution failure. Next, it is affected by the incomplete term that the total mass matrix contains negative elements, which are not located in the main diagonal of the matrix. This is in contradiction with the positive constant of the linear element mass matrix and also has no physical meaning. Consequently, it may increase the disparity between the FEM results and correct solutions. If a linear element is adopted and the enriched ridge shape function Eq. (9.29) is used to simulate two-phase flows, Sauerland and Fries (2011) also pointed out that the iteration steps are dramatically increased and the solution cannot converge. This situation is similar to using Eq. (9.26). It is thus clear that Eqs. (9.26) and (9.29) are not suitable for use in transient problems with weak discontinuity in velocity.

The bubble function proposed by Minev et al. (2003) also contains incomplete high-order terms, and is not suitable for use with linear elements. The enriched shape function proposed by Coppola-Owen and Codina (2005) is continuous on both sides of the interface in the element. To guarantee continuity and locality, this shape function is continuous at the interface and zero at the node. Therefore, it has a particular gradient in the element and discontinuity at the border between the elements. This characteristic of the interpolation function is suitable for use in a pressure discrete solution, because the pressure in the N-S equation only requires L^2 continuity. However, this kind of discontinuity is not suitable for velocity discrete solutions, because the derivative of velocity with respect to spatial coordinate is involved in the N-S equation. The finite element domain of velocity must have H^1 continuity.

In fact, the work of Smolianski (2005) and Tornberg and Engquist (2000) reminds us that although the differential significance in the FEM solution cannot precisely satisfy the weak discontinuity condition, error is introduced. However, the precise element integral may remedy this shortcoming. Consequently, it is not necessary to provide special treatment for the velocity u in the solution process. If the pressure distribution of the fluid field is described precisely near the interface, the discrete pressure field, Eq. (9.30), must be adopted, which has been proved by the works of Sauerland and Fries (2011) and Mier-Torrecilla et al. (2010). In the examples we have given, Eq. (9.30) is used to simulate pressure discontinuity.

The precise integral method and the convention shape function are used for velocity interpolation.

In addition, in the finite element implementation of the N-S equation, all of the elements are linear elements, and the high order terms, as for the divergence term of viscosity stress, cannot be obtained by nodal value interpolation. Thus, L^2 projection is used. The primary differential equation is transformed to a convective term and a pressure term, as well as its projection, so the algorithm's compatibility is guaranteed. It is clear that there is a difference compared with the case of single-phase flow. The convective term and pressure gradient term in two-phase flows contribute to discontinuity at the interface; thus, the enriched shape function is applied in their projection to avoid the additional numerical difficulty caused by the discontinuity.

9.4 DISCRETIZATION OF GOVERNING EQUATIONS

After the interpolation method is determined, finite element discretization for the N-S equation and level set evolution equation can be carried out.

Since the orientation is included in the convective term, the numerical noise may be simulated by the isotropic version of the standard Galerkin method. Thus, the Streamline Upwind Petrov Galerkin (SUPG) method is adopted to deal with the fluid field momentum equation consisting of the convective term and the level set evolution equation. In the time domain, the backward Euler format is used; in a spatial domain, a triangular element is adopted to discretize the computational domain. In order to get a convenient computational scheme, linear polynomial shape functions are used for velocity and pressure. However, since we have an incompressible fluid, the standard Galerkin method may result in pressure oscillation. Therefore, the Pressure Stabilizing Petrov Galerkin (PSPG) method is used to deal with pressure stabilization.

The N-S equation is discretized in the spatial domain, such that

$$u^{n+1} = \left[\left[N_1 \quad \cdots \quad N_{n_{\text{node}}}\right]\left[u_1^{n+1} \quad \cdots \quad u_{n_{\text{node}}}^{n+1}\right]^{\text{T}} = [N]^{\text{T}}\left[u^{n+1}\right]\right. \tag{9.32}$$

$$u^n = \left[N_1 \quad \cdots \quad N_{n_{\text{node}}}\right]\left[u_1^n \quad \cdots \quad u_{n_{\text{node}}}^n\right]^{\text{T}} = [N]^{\text{T}}[u^n] \tag{9.33}$$

Similarly,

$$p = \left[N_1 \quad \cdots \quad N_{n_{\text{node}}} \quad N_1^{\text{enr}} \quad \cdots \quad N_{n_{\text{enr}}}^{\text{enr}}\right]\left[p_1 \quad \cdots \quad p_{n_{\text{node}}} \quad p_1^{\text{enr}} \quad \cdots \quad p_{n_{\text{enr}}}^{\text{enr}}\right]^{\text{T}}$$
$$= \left[\overline{N}\right]^{\text{T}}[p]$$

$$\tag{9.34}$$

It is noted that the difference between $[N]$ and $[\overline{N}]$ lies only in the enriched shape function. In the following discussion, all the square brackets are omitted for convenience. The time discretization of the N-S equation is carried out.

The time increment is taken as $\Delta t = t^{n+1} - t^n$ and boundary conditions substituted, then the set of equation are discretized:

$$\frac{1}{\Delta t}M_u\left(u^{n+1} - u^n\right) + C_u u^{n+1} + K_u u^{n+1} - G_p p - F_{\text{grav}} - F_{\text{ST}} - F_{\text{ext}}$$

$$+K_{\text{SUPG}} u^{n+1} + K_{\text{conv}} \Pi_{\text{conv}} + K_{\text{LSIC}} u^{n+1} = 0 \tag{9.35}$$

$$D_u u^{n+1} + K_{\text{PSPG}} p + K_{\text{PGP}} \Pi_p = 0 \tag{9.36}$$

$$C_{\text{conv}} u^{n+1} + M_{\text{conv}} \Pi_{\text{conv}} = 0 \tag{9.37}$$

$$G_{\text{PGP}} p + M_{\text{PGP}} \Pi_p = 0 \tag{9.38}$$

where

$$M_u = \int_\Omega \rho^{n+1} N N^{\text{T}} d\Omega \qquad C_u = \int_\Omega \rho^{n+1} N u^{n+1} \cdot \nabla N^{\text{T}} d\Omega$$

$$K_u = \int_\Omega \mu^{n+1}(SN) I_0 (SN)^{\text{T}} d\Omega \quad G_p = \int_\Omega (\nabla N) \overline{N}^{\text{T}} d\Omega$$

$$F_{\text{grav}} = \int_\Omega \rho^{n+1} N g d\Omega \quad F_{\text{ST}} = \int_{\gamma_{\text{int}}} N \gamma \kappa n d\gamma \quad F_{\text{ext}} = \int_{\gamma_t} N \overline{t} d\gamma$$

$$K_{\text{SUPG}} = \sum_{e=1}^{n_{elem}} \int_{\Omega^e} \tau_{\text{SUPG}} (u^{n+1} \cdot \nabla N)(\rho^{n+1} u^{n+1} \cdot \nabla N)^{\text{T}} d\Omega \tag{9.39}$$

$$K_{\text{conv}} = \sum_{e=1}^{n_{elem}} \int_{\Omega^e} \tau_{\text{SUPG}} (u^{n+1} \cdot \nabla N) N_{\text{conv}}^{\text{T}} d\Omega$$

$$K_{\text{LSIC}} = \sum_{e=1}^{n_{elem}} \int_{\Omega^e} \rho^{n+1} \tau_{\text{LSIC}} (\nabla \cdot N)(\nabla \cdot N)^{\text{T}} d\Omega$$

$$D_u = \int_\Omega \overline{N}(\nabla N)^{\text{T}} d\Omega = G_p^{\text{T}}$$

$$K_{\text{PSPG}} = \sum_{e=1}^{n_{elem}} \int_{\Omega^e} \frac{1}{\rho^{n+1}} \tau_{\text{PSPG}} \nabla \overline{N} \cdot \nabla \overline{N}^{\text{T}} d\Omega \tag{9.40}$$

$$K_{\text{PGP}} = \sum_{e=1}^{n_{elem}} \int_{\Omega^e} \frac{1}{\rho^{n+1}} \tau_{\text{PSPG}} \nabla \overline{N} \cdot \overline{N}^{\text{T}} d\Omega$$

$$C_{\text{conv}} = \int_\Omega \rho^{n+1} \overline{N}(\rho^{n+1} u^{n+1} \cdot \nabla N)^{\text{T}} d\Omega$$

$$M_{\text{conv}} = \int_\Omega \rho^{n+1} \overline{N}\, \overline{N}^{\text{T}} d\Omega \tag{9.41}$$

$$G_{\mathrm{PGP}} = \sum_{e=1}^{n_{elem}} \int_{\Omega^e} \frac{1}{\rho^{n+1}} \tau_{\mathrm{PSPG}} \overline{N} \cdot \nabla \overline{N}^{\mathrm{T}} \mathrm{d}\Omega = K_{\mathrm{PGP}}^{\mathrm{T}}$$

(9.42)

$$M_{\mathrm{PGP}} = \sum_{e=1}^{n_{elem}} \int_{\Omega^e} \frac{1}{\rho^{n+1}} \tau_{\mathrm{PSPG}} \overline{N}\, \overline{N}^{\mathrm{T}} \mathrm{d}\Omega$$

and

$$S = \begin{bmatrix} \dfrac{\partial}{\partial x} & 0 & \dfrac{\partial}{\partial y} \\[2mm] 0 & \dfrac{\partial}{\partial y} & \dfrac{\partial}{\partial x} \end{bmatrix}^{\mathrm{T}}$$

(9.43)

$$I_0 = \begin{bmatrix} 2 & 0 & 0 \\ 0 & 2 & 0 \\ 0 & 0 & 1 \end{bmatrix}$$

(9.44)

The superscripts in ρ^{n+1} and u^{n+1} indicate that the spatial distributions of both quantities are determined by the interfacial position at time t^{n+1}. τ_{SUPG}, τ_{PSPG}, and τ_{LSIC} are defined as the stabilized coefficients corresponding to each element, in which

$$\tau_{\mathrm{SUPG}} = \left[\left(\frac{2\|u\|}{h^e} \right)^2 + \left(\frac{4u}{\rho(h^e)^2} \right)^2 \right]^{-1/2}$$

(9.45)

$$\tau_{\mathrm{PSPG}} = \tau_{\mathrm{SUPG}}$$

(9.46)

$$\tau_{\mathrm{LSIC}} = \frac{1}{2} \alpha^e h^e \|u\|$$

(9.47)

where h^e represents the characteristic size of the element, which can be understood as the projection of the element along the flow velocity direction. α^e is a numerical parameter related to element Reynolds number Re^e, which can be written as

$$\alpha^e = \min\left(\frac{\mathrm{Re}^e}{3}, 1 \right)$$

(9.48)

while the element Reynolds number is

$$\mathrm{Re}^e = \frac{\rho \|u\| h^e}{2\mu}$$

(9.49)

It is worth noting that τ_{SUPG} and τ_{PSPG} have the dimension of time, which can be regarded as the stable time based on the SUPG and PSPG algorithms.

The divergence integral term corresponding to τ_{LSIC} is the weighting approximation of velocity divergence in the shape function divergence spatial domain. When the velocity field satisfies the mass conservation equation (9.1) precisely, this term disappears. Therefore, this term can be regarded as compensation of the mass conservation condition affected by the stabilized terms of SUPG and PSPG. With respect to the details of the algorithms SUPG and PSPG, see Shakib et al. (1991) and Tezduyar and Osawa (2000).

It must be noticed that Eq. (9. 37) defines the weighting projection Π_{conv} of the convective term with opposite sign. The application of this projection term is considered to provide numerical algorithm compatibility. In fact, the two terms for τ_{SUPG} in Eq. (9.35) may be combined to give

$$
\boldsymbol{K}_{\text{SUPG}}\boldsymbol{u}^{n+1} + \boldsymbol{K}_{\text{conv}}\Pi_{\text{conv}}
$$

$$
= \sum_{e=1}^{n_{\text{elem}}} \int_{\Omega^e} \tau_{\text{SUPG}} (\boldsymbol{u}^{n+1} \cdot \nabla N) \left(\rho \left(\frac{\partial \boldsymbol{u}}{\partial t} + \boldsymbol{u} \cdot \nabla \boldsymbol{u} \right) - \nabla \cdot (\mu(\nabla \boldsymbol{u} + \boldsymbol{u}\nabla)) \right.
$$

$$
\left. + \nabla p - \rho \boldsymbol{f} - \gamma \kappa \delta(\boldsymbol{x} - \bar{\boldsymbol{x}})\boldsymbol{n} \right) \text{d}\Omega \tag{9.50}
$$

However, because the linear element cannot represent high-order spatial derivative terms, like viscosity stress, the terms on the right-hand side of Eq. (9.50) become

$$
\sum_{e=1}^{n_{\text{elem}}} \int_{\Omega^e} \tau_{\text{SUPG}} (\boldsymbol{u}^{n+1} \cdot \nabla N) \left(\rho \left(\frac{\partial \boldsymbol{u}}{\partial t} + \boldsymbol{u} \cdot \nabla \boldsymbol{u} \right) + \nabla p - \rho \boldsymbol{f} - \gamma \kappa \delta(\boldsymbol{x} - \bar{\boldsymbol{x}})\boldsymbol{n} \right) \text{d}\Omega
$$

$$
\tag{9.51}
$$

When the mesh size tends to zero, the finite element results do not approach accurate solutions. In other words, this algorithm is incompatible. Based on the concept of Oñate et al. (2006) and considering momentum equation (9.10), it is understood that the convective term $\rho \boldsymbol{u} \cdot \nabla \boldsymbol{u}$ is a spatial term, which has no explicit relation with time and can be represented by a linear shape function. Therefore, the momentum equation (9.10) can be written in a form corresponding to the convective term, such that

$$
\rho \boldsymbol{u} \cdot \nabla \boldsymbol{u} + \Pi_{\text{conv}} = 0 \tag{9.52}
$$

where

$$
\Pi_{\text{conv}} = \rho \frac{\partial \boldsymbol{u}}{\partial t} - \nabla \cdot (\mu(\nabla \boldsymbol{u} + \boldsymbol{u}\nabla)) + \nabla p - \rho \boldsymbol{f} - \gamma \kappa \delta(\boldsymbol{x} - \bar{\boldsymbol{x}})\boldsymbol{n} \tag{9.53}
$$

It is seen that the density ρ is discontinuous at the interface, but it is usually difficult to cancel out this discontinuity from the $\boldsymbol{u} \cdot \nabla \boldsymbol{u}$ term. Therefore, the convective term is discontinuous at the interface. When finite element discretization is carried out for Eq. (9.52), the weighting function $\rho^{n+1}\overline{N}$ is selected to be able to characterize the interface discontinuity. Thus, the SUPG term in Eqs. (9.35) and (9.37) can be obtained, where $\overline{N}$ is the same shape function, to characterize the pressure discontinuity. It can be seen by considering M_{conv} that, in single-phase flow, we have the same M_{conv} and M_u because we have the same $\overline{N}$ and N. However, in two-phase flows, $\overline{N}$ is an extended term, so the two matrices are not equal to each other.

For the same reason, considering the compatibility of mass conservation equation (9.1), a similar treatment can be carried out for the pressure gradient; thus, the momentum equation (9.10) can be written as

$$\nabla p + \Pi_p = 0 \tag{9.54}$$

where

$$\Pi_p = \rho\left(\frac{\partial u}{\partial t} + u \cdot \nabla u\right) - \nabla \cdot (\mu(\nabla u + u\nabla)) - \rho f - \gamma \kappa \delta(x - \overline{x})n \tag{9.55}$$

Since ∇p shows a discontinuity at the interface, the weighting function $(1/\rho)\tau_{\text{PSPG}}\overline{N}$ is used in the weighting integral. Thus, the PSPG term in Eqs. (9.36) and (9.38) can be obtained.

For the level set enriched equation (9.17), we have

$$\phi^{n+1} = [N_1 \quad \cdots \quad N_{n_{\text{node}}}][\phi_1^{n+1} \quad \cdots \quad \phi_{n_{\text{node}}}^{n+1}]^{\text{T}} = [N]^{\text{T}}[\phi^{n+1}] \tag{9.56}$$

$$\phi^n = [N_1 \quad \cdots \quad N_{n_{\text{node}}}][\phi_1^n \quad \cdots \quad \phi_{n_{\text{node}}}^n]^{\text{T}} = [N]^{\text{T}}[\phi^n] \tag{9.57}$$

In the same way, the SUPG method and backwards Euler format are used for discretization. The discretized equation is given by

$$\frac{1}{\Delta t}M_\phi(\phi^{n+1} - \phi^n) + C_\phi \phi^{n+1} + F_\phi = 0 \tag{9.58}$$

where

$$M_\phi = \int_\Omega NN^{\text{T}}\mathrm{d}\Omega$$

$$C_\phi = \int_\Omega Nu^{n+1} \cdot \nabla N^{\text{T}}\mathrm{d}\Omega \tag{9.59}$$

$$F_\phi = \sum_{e=1}^{n_{elem}} \int_{\Omega^e} \tau_\phi(u^{n+1} \cdot \nabla N)\left(\frac{1}{\Delta t}(\phi^{n+1} - \phi^n) + u^{n+1} \cdot \nabla\phi\right)\mathrm{d}\Omega$$

The stability coefficient τ_ϕ is

$$\tau_\phi = \frac{h^e}{2\|u\|} \tag{9.60}$$

In this way, the governing equations have been discretized.

9.5 NUMERICAL INTEGRAL METHOD

As discussed previously, it is important to describe the fluid field near the interface precisely. It is necessary to consider the interfacial factor in the numerical integral scheme when the total stiffness matrix is assembled. As in many references on X-FEM, since the linear triangular element is used, the interface becomes one of segment lines in accordance with the interfacial position. The element is repartitioned into sub-elements as shown in Figure 9.4, and then the integral is processed in the sub-element. It is seen that the objective of the repartitioned element is only precise integration, with the aim of capturing the field information precisely near the interface, and it does not introduce new nodes or new DOFs.

We know that the calculation of surface tension is involved in momentum equation (9.35), while the surface tension is related to the interfacial curvature; thus, it is necessary to obtain the value of curvature. The relation between surface curvature and surface normal is satisfied:

$$\kappa = -\nabla \cdot \boldsymbol{n} \qquad (9.61)$$

while the surface normal vector can be expressed by the gradient of the level set, such that

$$\boldsymbol{n} = \frac{\nabla \phi}{\|\nabla \phi\|} \qquad (9.62)$$

Therefore,

$$\kappa = -\nabla \cdot \left(\frac{\nabla \phi}{\|\nabla \phi\|} \right) \qquad (9.63)$$

However, the linear triangular element is a constant strain element, which cannot directly express more than second-order spatial derivative. As a result, it is necessary to deal with the gradient $\nabla \phi$ of the level set. From the work of Chessa and Belytschko (2003b), as well as Smolianski (2001), the L^2 projection is used via

FIGURE 9.4   Integral scheme in sub-element.

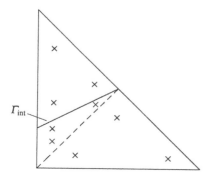

$$M_n \boldsymbol{n}^{\mathrm{h}} = \boldsymbol{F}_n \tag{9.64}$$

to solve $\boldsymbol{n}^{\mathrm{h}}$, where

$$
\begin{aligned}
M_n &= \int_\Omega N N^{\mathrm{T}} \mathrm{d}\Omega \\[2mm]
F_n &= \int_\Omega N \frac{\nabla \phi}{\|\nabla \phi\|} \mathrm{d}\Omega
\end{aligned}
\tag{9.65}
$$

Then,

$$M_\kappa \kappa^{\mathrm{h}} = -\boldsymbol{F}_\kappa + G_n \boldsymbol{n}^{\mathrm{h}} \tag{9.66}$$

is used to find the approximate value of curvature κ^{h}, where

$$
\begin{aligned}
M_\kappa &= \int_\Omega N N^{\mathrm{T}} \mathrm{d}\Omega \\[2mm]
F_\kappa &= \int_{\partial\Omega} \bar{\boldsymbol{n}} \cdot \boldsymbol{n}^{\mathrm{h}} N \mathrm{d}\gamma \\[2mm]
G_n &= \int_\Omega (\nabla N) \cdot N^{\mathrm{T}} \mathrm{d}\Omega
\end{aligned}
\tag{9.67}
$$

In the expression for F_k, $\partial\Omega$ indicates the outside boundary of the total calculation domain and $\bar{n}$ is the outer normal direction vector.

The unified solution can be taken to solve a set of discretized N-S equations (9.35)–(9.38). It is combined with these equations to solve u, p, Π_{conv},

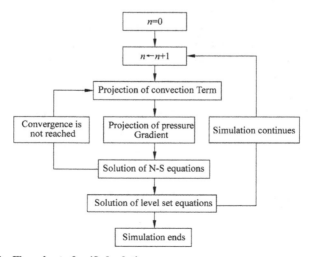

FIGURE 9.5 Flow chart of unified solution process.

and Π_p simultaneously. An alternative is to use an iteration method to solve these equations in turn. In order to reduce the computational requirements, the latter method is used here. The corresponding relation is set up explicitly between interfacial evolution and the N-S equation. A flow chart of the unified solution process is illustrated in Figure 9.5.

9.6 EXAMPLES AND ANALYSES

The problem considered here is for a single bubble rising in water due to buoyancy. It is assumed that the flow is incompressible as the Mach number is less than 0.3. The simulation method is verified using this numerical example. A circular bubble with diameter 0.5 m is initially placed in a rectangular domain with coordinate range $[-0.5, 0.5] \times [0, 2]$, as shown in Figure 9.6. It is supposed there are no sliding boundaries on any of the borders. On the top surface, the pressure reference value is set to be 0.0. The initial position of the center of a circle bubble is $(0, 0.5)$. Starting at time $t = 0$, the water and bubble suspended inside are in a steady state. The computational mesh with 2373 nodes and 4542 linear triangular elements is shown in Figure 9.7. The initially quiescent bubble starts to rise due to buoyancy from the surrounding fluid at the beginning. In this process, the fluidic and interfacial properties determine the subsequent deformation of the bubble.

The flow state is dependent on four dimensionless parameters, which are ratio of density ρ_1/ρ_2, ratio of viscosity coefficient μ_1/μ_2, Eötvös number $Eo = 4(\rho_1 - \rho_2)gR^2/\gamma$, and Morton number $M = g\mu_1^4(\rho_1 - \rho_2)/(\rho_1^2\gamma^3)$. In the computation, the liquid density, ratio of density, ratio of viscosity coefficient, and Morton number are set to be $\rho_1 = 1000$, $\rho_1/\rho_2 = 1000$, $\mu_1/\mu_2 = 100$,

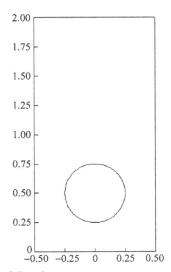

FIGURE 9.6   Computational domain.

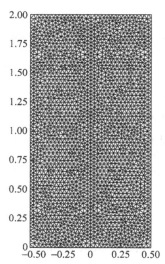

FIGURE 9.7 Computational mesh.

and $M = 0.01$ respectively. Taking $Eo = (1,10,100,1000)$, the flow state is investigated.

The computation results are shown in Figure 9.8 with the diverse shapes of bubble corresponding to various Eo numbers, as well as streamlines of fluid field. When the Eo numbers are set to 1, 10, 100, and 1000 respectively, the configurations of the bubble correspond to spherical, ellipsoidal, dimpled ellipsoidal-cap and skirted, which can be compared well with experimental data and theoretical analyses provided by Clift et al. (1978). It can be seen that the deformation capacity of the bubble gradually increases as Eo number

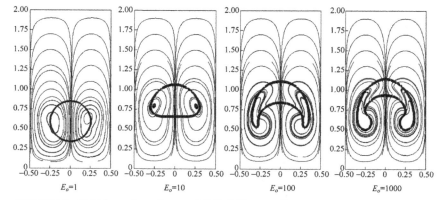

FIGURE 9.8 Bubble configurations and fluid field streamlines.

increases. In reality, the larger the *Eo* numbers, the smaller the effect of surface tension on flow. The fluid motion becomes more free near the interface, especially in the wake area.

In the case of *Eo* = 10, the variations in configuration of the bubble during the rising process are plotted in Figure 9.9. It can be seen that the velocity in the lower part of bubble is much faster, which is influenced by the wake. The bubble rises much faster and is compressed in the vertical direction. However, the velocity in the lower part of bubble is restricted by the surface tension and cannot continue to match the velocity of the upper part. Therefore, the final configuration of the bubble is elliptical on the upper part and flat on the lower part. In the same situation, the pressure distribution of the fluid field is plotted in Figure 9.10. It can be seen that the internal and external pressure gradients of the bubble show great disparity, which results in relatively large surface tension.

In the case of *Eo* = 1000, there is no significant influence on the flow by surface tension, as shown in Figure 9.11. Due to the effect of the wake, the lower part of the bubble soon catches up with the upper part. This indicates that the surface tension does not significantly restrict bubble deformation. Part of the bubble subsequently has a separate configuration. This phenomenon can be understood since the original bubble lacks sufficient surface tension to restrict deformation, so its energy is reduced as it is used for bubble deformation; after the separate configuration occurs, a portion of the bubble's energy is taken by the separate parts, thus the total energy is reduced. This tendency for stability results in the deformation occurring more easily.

Figure 9.12 illustrates the pressure distributions at different times for *Eo* = 1000. The situation differs from the case of *Eo* = 10. The pressure field here has no great disparity in discontinuity at the interface, because of the relatively smaller surface tension.

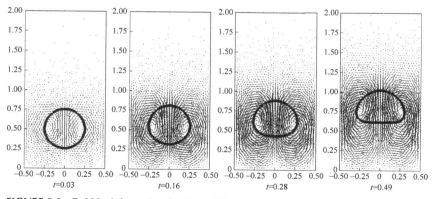

FIGURE 9.9   Bubble deformation for *Eo* = 10.

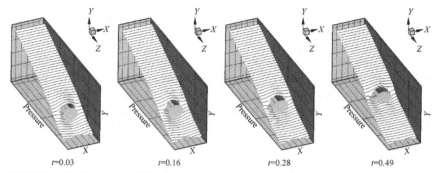

FIGURE 9.10 Pressure field for $Eo = 10$.

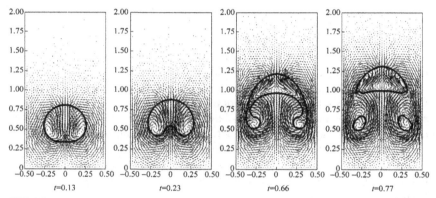

FIGURE 9.11 Bubble deformation for $Eo = 1000$.

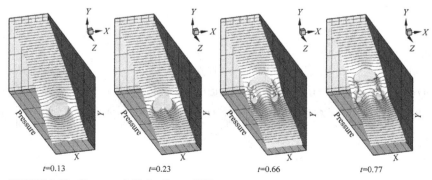

FIGURE 9.12 Pressure field for $Eo = 1000$.

9.7 SUMMARY

Based on the X-FEM concept, a computational method is proposed in this chapter, which can be used to express the discontinuity field of two-phase flows in a coarse mesh. In particular, the projection term consisting of discontinuity is introduced in solving the N-S equation. Consequently, the potential problem of SUPG and PSPG algorithm incompatibility can be overcome. In addition, it can be observed from computations that the weak discontinuity in velocity may influence the accuracy of the solution, and may even result in solution failure. This indirectly highlights the importance of precise integration. By comparison of the results with those of Smolianski (2005), Clift et al. (1978) and Fries (2008), it can be concluded that the method is particularly suitable for use in the numerical simulation of transient immiscible and incompressible two-phase flows.

Chapter 10

Research Progress and Challenges of X-FEM

Chapter Outline

10.1 Research on Micro-Scale Crystal Plasticity 213

  10.1.1 Discrete Dislocation Plasticity Modeling   215

  10.1.2 X-FEM Simulation of Dislocations   219

10.2 Application of Multi-Scale Simulation   223

10.3 Modeling of Deformation Localization   224

10.4 Summary   228

10.1 RESEARCH ON MICRO-SCALE CRYSTAL PLASTICITY

Among the diverse applications of micro-devices in recent years, the control of plastic formation during micromachining of microelectromechanical systems (MEMS) and the prediction of mechanical reliability in service have become hot topics in the study of material science and mechanics. The plastic deformation of crystalline materials at the submicron scale ranging from hundreds of nanometers to tens of micrometers exhibits many new characteristics: (1) A strong size effect of yield stress appears even under uniform loading; (2) the plastic flow proceeds in a strongly temporal intermittent and spatially localized manner, in which slip lines or slip bands are observed at the surface of deformed crystals; (3) because of the space limitation, the mechanism that dominates is dislocation storage and multiplication change, leading to new stress—strain relations (Liu et al., 2009a). These new characteristics present a great challenge to the research in this area. The stress—strain curves in micropillar compression tests are characterized by serrated yielding (Csikor et al., 2007) as shown in Figure 10.1, and fine discrete slip bands can be observed along the gauge length of the samples associated with dislocation avalanches, as shown in Figure 10.2.

Although direct large-scale atomistic calculations have provided a number of meaningful results (Abraham et al., 1997; Bachlechner et al., 1998; Zhou

Extended Finite Element Method. http://dx.doi.org/10.1016/B978-0-12-407717-1.00010-8

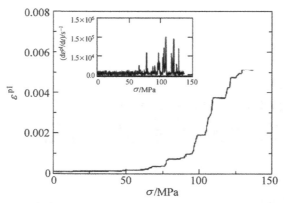

FIGURE 10.1 A typical stress—strain curve obtained from simulation of the three-dimensional dislocation dynamics model in a load controlled test (Csikor et al., 2007).

et al., 1998), the problems that these atomistic simulations can handle are still limited by the computational ability. For example, the number of atoms that a typical molecular dynamics (MD) simulation can handle is about 10^8, and the time step is about 10^{-14} s. Using MD to simulate the plastic deformation of a cubic crystal with a side length of 10 μm, about 10^{13} atoms and 10^8 time steps are required. This computational scale is far beyond the capabilities of current computers. On the other hand, in current MD simulations, usually an extremely high loading rate is used to speed the calculation. However, this high loading rate may influence the dynamic behavior of dislocation, which is the basic carrier of plastic strain in metals, and the results obtained under such conditions are questionable (Needleman, 1999). Because of the limitation of computational ability in discrete simulation, it is very useful to apply a

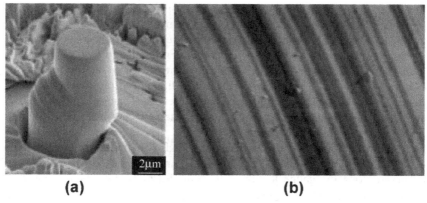

(a)   **(b)**

FIGURE 10.2 Slip band (a; Uchic et al., 2004) and slip line (b; Neuhauser, 1983) formed on the surface of crystals during compression loading.

continuum method to solve the plastic deformation at the submicron scale. Therefore, in the following subsections, two computational methods under the continuum mechanics framework are briefly introduced.

10.1.1 Discrete Dislocation Plasticity Modeling

In discrete dislocation theory, the dislocation is considered as the essential carrier of plastic flow. The plastic deformation originates from the quantities of defects, so-called "dislocation", in the crystal, as shown in Figure 10.3, where **b** denotes the Burgers vector. By dislocation slip, plastic deformation in crystals can occur under relatively low stress without the necessity of moving two neighboring atomic planes. In discrete dislocation plasticity, the dislocations are modeled as line defects embedded in the elastic media. The dynamic calculations for a number of interacting dislocations was not realized until the 1980s. One of the most important achievements in this field is that a 3D discrete dislocation dynamics (3D-DDD) code was developed that includes the effect of crystal orientation and slip geometry (Kubin and Canova, 1992). By 3D-DDD simulation, we can predict the plastic deformation of a crystalline material in micro-scale, and further get the mechanical properties of crystals. The 3D-DDD method provides a powerful tool for studying the plastic behavior of crystals at the submicron level (Liu et al., 2009b; Gao et al., 2011).

Recently a computational model for discrete dislocation plasticity completely based on continuum mechanics was developed by the authors by directly combining 3D discrete dislocation dynamics (DDD) and the finite element method (FEM) (Liu et al., 2009b). In this model, the discrete dislocation plasticity in the micro-scale crystal is solved completely under a continuum mechanics framework: (1) an initial internal stress field is introduced to represent the pre-existing stationary dislocations in the crystal; (2) the external boundary condition is handled spontaneously using the finite element method; (3) the constitutive relationship is based on the finite deformation theory of crystal plasticity, but the discrete plastic strains are calculated by dislocation dynamics methodology. The information transfer between 3D-DDD and FEM is shown in Figure 10.4. When considering the lattice of a

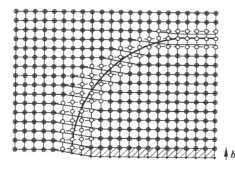

FIGURE 10.3 Dislocation — the line defect in a crystal.

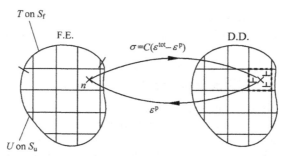

FIGURE 10.4 Schematic of the stress and plastic strain transfer between dislocation dynamics code and finite element code.

crystal as the element in FEM, as in the crack propagation problem, when the dislocation slips in the lattice, it induces a displacement discontinuity across the slip plane. So the concept of X-FEM can be applied to solve the dislocation problem.

In the computational model, the newly nucleated or propagating dislocations are represented by discrete plastic strains induced by their slip. Then these discrete plastic strains are localized to the continuum material points to introduce the stress field associated with newly nucleated or propagating dislocations. In the work of Zbib and Diaz de la Rubia (2002), the plastic strains induced by a dislocation segment motion are all localized to the continuum point (FEM) that it belongs to. In practice, it is difficult to capture the stress field near the dislocation line using this method (Zbib and Diaz de la Rubia, 2002), which is important at the submicron-to-nanometer scale. On the other hand, this localization method will induce strong discontinuity of the plastic strain field between different elements because of the discrete nature of dislocation motion. This discontinuity will further disturb the FEM solution.

In the classical dislocation theory model, the energy and stress can be infinite unless some truncation scheme is applied. This singularity originates from an unphysical but mathematically convenient description of the dislocation core in which the Burgers vector is described by a delta function. In our computational model, the finite dislocation core effect is taken into account, in which the dislocation is no longer treated as an infinitely thin filament. Here a Burgers vector density function $\mathbf{g}(\mathbf{x})$ is introduced to remove the dislocation singularity by spreading its Burgers vector $\mathbf{b}$ among every point around the dislocation line, assuming that dislocation lines (like cracks) pass several lattice elements. The Burgers vector $\mathbf{b}$ is expressed as follows:

$$\mathbf{b} = \int \mathbf{g}(\mathbf{x}) \mathrm{d}^3 \mathbf{x} = \int \mathbf{b}\omega(\mathbf{x}) \mathrm{d}^3 \mathbf{x} \tag{10.1}$$

where $\omega(\mathbf{x})$ is the Burgers vector distribution function. The formulation of $\omega(\mathbf{x})$ is rather general and can be used to account for various realistic details of the dislocation core structure, e.g., possible splitting of a perfect dislocation into partial dislocation in FCC crystals. Cai et al. (2006) have developed an isotropic Burgers vector distribution function to remove the intrinsic singularity of classical continuum theory. Here we introduce a simpler isotropic Burgers vector distribution function $\omega(\mathbf{x})$ as follows, which is often used in the Peierls–Nabarro model of dislocations:

$$\omega(\mathbf{x}) = \omega(r) = \frac{1}{\pi}\frac{a}{r^2 + a^2}, \quad r = ||\mathbf{x}|| \tag{10.2}$$

where a is the spreading radius, which can vary with different element size, integration scheme, and the precision of the problems considered. Indeed, a proper description of any intrinsic length scale, like the spreading radius in our model, correspondingly requires an adequate refinement of the FEM mesh. The Burgers vector distribution $\omega(\mathbf{x})$ is strongly nonlinear around the dislocation line, as shown in Figure 10.5.

So the plastic strain can be localized to the continuum material points (integration points in FEM) around the dislocation segments, e.g., the shaded elements in Figure 10.6, by a simple isotropic Burgers vector density function described above. In our model the plastic strain $\boldsymbol{\varepsilon}_p^{(i)}$ at each material point (integration point) i can be obtained as follows:

$$\boldsymbol{\varepsilon}_p^{(i)} = \frac{w^{(i)}}{\sum_{i=1}^{n} w^{(i)}}\boldsymbol{\varepsilon}_p \tag{10.3}$$

where $w^{(i)}$ is the weight function of each material point i, ε_p is the total plastic strain induced by the slip of dislocation segments, and the number n of considered continuum material points in this homogenization process is decided by the precision of the problems considered. The weight function $w^{(i)}$

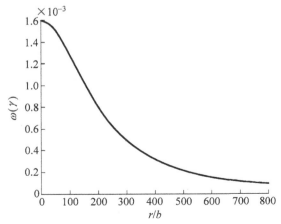

FIGURE 10.5   Plot of Burgers vector distribution function (spreading radius $a = 200b$; b is the magnitude of Burgers vector).

FIGURE 10.6 When a dislocation line slips a small distance l in an element, the displacement discontinuity induced is localized in the shaded elements.

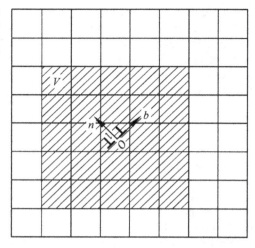

of material point i in the interpolation is related to the total Burgers vector distribution in the material point i, and it can be expressed as

$$w^{(i)} = \int_{V^i} \omega(\mathbf{x}) dV \tag{10.4}$$

where V^i is the volume occupied by material point i. Equation (10.4) can be integrated numerically. Linear elements with eight nodes and eight integration points are used in the calculation.

At each time increment, the slip distance Δl of each dislocation segment is very small compared with the length of dislocation segments, so the slipped area in each time increment can be regarded as a "dislocation pole", as shown in Figure 10.6. We establish a local coordinate system at the center of the "dislocation pole", shown in Figure 10.6, where $\mathbf{n}$ is the normal direction of the slip plane, $\mathbf{b}$ is the Burgers vector direction, and ξ is the dislocation line direction. Then Eq. (10.4) is numerically integrated in this local coordinate system. To improve efficiency, we introduce three cutoff distances in these three axis directions. If the local coordinates of integration points exceed these cutoff distances, their weight functions will not be considered.

The spreading radius a and three cutoff parameters can be chosen to approximately match the energy dissipated in dislocation dynamics and FEM:

$$F \cdot \Delta l = \sum_{i=1}^{n} \int_{V^i} \sigma_{equ}^{(i)} \cdot \Delta \varepsilon_{equ}^{(i)} dV \tag{10.5}$$

where F is the total force acting on the dislocation line, Δl is the slip distance at each time increment, and $\sigma_{equ}^{(i)}$ and $\Delta \varepsilon_{equ}^{(i)}$ are equivalent stress and equivalent plastic strain increments of material point i respectively.

The localization method in our model has clear physical sense, and can easily be implemented into 3D finite element code. It works well in handling crystal material of different crystallographies, as can be seen in the following examples. Next numerical simulations are carried out to prove the capability of the computational model in predicting the plastic behaviors of crystals at submicron-to-nanometer scales.

An isotropic copper single crystal is considered with the following properties (relevant to both the continuum model and DDD model): shear modulus $\mu = 42$ GPa, Poisson's ratio $\nu = 0.347$, density $\rho = 8900$ kg/m^3, magnitude of Burgers vector $b = 0.256$ nm, and static viscous drag coefficient $B_0 = 5.5 \times 10^{-5}$ Pa s^{-1} at a temperature of 300 K. The (x, y, z) coordinates show the crystallographic orientations. The element type in the simulations is an eight-node brick element with eight integration points. The time increment is $10^{-11}-10^{-10}$ s. In the simulation, we assume a dislocation line with slip system $(011)/[01\bar{1}]$ propagated after nucleation from one edge of the top surface of the crystal; one corner of the crystal is fixed to eliminate the rigid body displacement. When the propagating dislocations arrive at the center of the crystal, the contours of the stress field and equivalent plastic strain are as shown in Figure 10.7.

As can be seen from the above simulation results, the discrete plastic strains induced by dislocation slip can be well localized to the continuum material points by spreading them around the dislocation line using the proposed Burgers vector density function. If the plastic strains associated with a moving dislocation segment are all localized to the element it belongs to (Zbib and Diaz de la Rubia, 2002), the stress and equivalent plastic strain distributions are as shown in Figure 10.8. As can be seen, the equivalent plastic strain is strongly discontinuous between neighboring elements. This discontinuity disturbs the results, and the self-stress field of the dislocation line cannot be well introduced by this localization method. Some extra work needs to be done to remedy this problem (Zbib and Diaz de la Rubia, 2002).

More complex processes, such as dislocation junction formation, loop dislocation evolution after nucleation from a free surface, etc., which could not be accommodated by research in the 2D context before, are also studied using this continuum-based discrete dislocation plasticity computational model (Liu et al., 2009b).

10.1.2 X-FEM Simulation of Dislocations

On the basis of crack simulation, Gracie et al. (2008) recently developed extended finite element method for the modeling of dislocations, providing a new way to investigate the micro-scale crystal plasticity under the continuum framework.

Like cracks, the geometry of dislocation can be described by a level set function. As shown in Figure 10.9(b), the dislocation line is characterized by

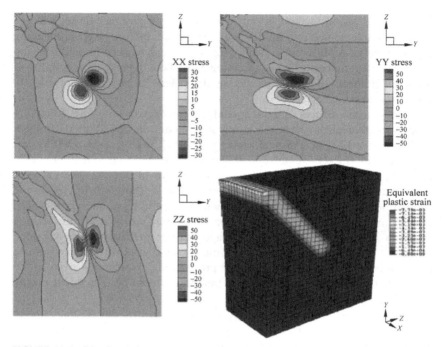

FIGURE 10.7 Distributions of stress (units MPa) and equivalent plastic strain around the dislocation core.

level set functions $f(\mathbf{x})$ and $g(\mathbf{x})$ respectively: the surface of the glide plane of dislocation is defined by $f(\mathbf{x}) = 0$; the core of the dislocation is then given by the intersection of the surface $f(\mathbf{x}) = 0$ and surface $g(\mathbf{x}) = 0$. The direction vector of dislocation is expressed as $\boldsymbol{\xi} = \nabla f \times \nabla g$.

Both dislocations and cracks are defects in the material and they have much in common. In the dislocation slipped region, the displacement across the glide plane is discontinuous and the singularity appears at the crack tip (as shown in Figure 10.9(a)). Similar to the enrichment for the crack surface and crack tip in Chapter 4, the displacement field in the crystal including n^{d} Volterra dislocations is expressed as

$$\mathbf{u}^{\mathrm{h}}(x) = \mathbf{u}^{\mathrm{FE}}(x) + \sum_{\alpha=1}^{n^{\mathrm{d}}} \mathbf{u}_\alpha^{\mathrm{D}}(x) \tag{10.6}$$

where $\mathbf{u}^{\mathrm{FE}}(x)$ is the standard continuous part of the displacement field and $\mathbf{u}_\alpha^{\mathrm{D}}(x)$ is the discontinuous or enriched part:

$$\mathbf{u}_\alpha^{\mathrm{D}}(x) = \mathbf{b}_\alpha \sum_{I \in S_\alpha^{\mathrm{jump}}} N_I(x)(H(f_\alpha(x)g_\alpha(x)) - H_{\alpha I})$$
$$+ b_\alpha \sum_{I \in S_\alpha^{\mathrm{core}}} N_I(x)(\mathbf{u}_\alpha^{\mathrm{core}}(f_\alpha(x), g_\alpha(x)) - \mathbf{u}_{\alpha I}^{\mathrm{core}}) \tag{10.7}$$

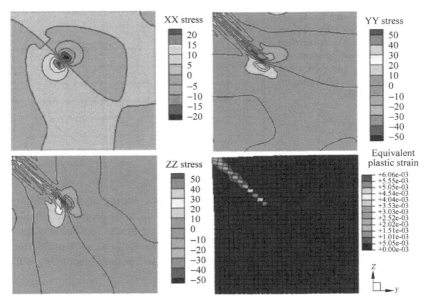

FIGURE 10.8   The stress (units MPa) and equivalent plastic strain distributions if the plastic strains are localized only to the element that the dislocation segment belongs to.

where the first term is Heaviside enrichment and $H(\bullet)$ is the standard Heaviside step function, $\mathbf{b}_\alpha$ is the Burgers vector of dislocation α, and node set S^{jump} denotes the nodes belonging to the elements with at least one edge crossed by the glide plane and far from the dislocation core. The second term is dislocation core enrichment node set S^{core}, which includes all the nodes around the dislocation core. One enrichment scheme is shown in Figure 10.10. In Eq. (10.7) the singular enrichment function $\mathbf{u}_\alpha^{\text{core}}$ is derived from the general solution of one dislocation in an infinite domain. It is also feasible to only use the

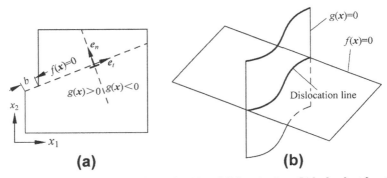

FIGURE 10.9   Definition of edge dislocation (a) and dislocation loop (b) by level set functions $f(\mathbf{x})$ and $g(\mathbf{x})$ (Gracie et al., 2008).

FIGURE 10.10 Schematic of enrichment for an edge disloca-tion (square nodes are enriched using the Heaviside function, cir-cular nodes are enriched using the dislocation core function) (Gracie et al., 2008).

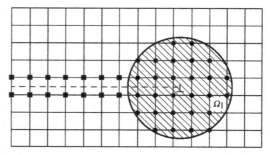

Heaviside enrichment function when it is difficult to get the analytical dislo-cation core enrichment function (for example, in anisotropic material); meanwhile, a fine mesh is needed to improve the accuracy and convergence rate. It should be noted that, compared with the formulation in crack problems, there are no additional degrees of freedom in the enriched displacement field for dislocation problems.

In the X-FEM simulation of dislocation dynamics, there are two methods to calculate the dislocation driving force, the so-called Peach—Koehler force. One is by using the J integral, which is the same as in fracture problems. The primary limitation of the contour integral approach is that the integral must be taken over a domain that does not contain any other dislocation cores (Gracie et al., 2008). It was found in Belytschko and Gracie (2007) that for cylindrical domains, a minimum internal radius of $3h_e$, where h_e is the average element size in the vicinity of the dislocation core, was required for accurate results. So to obtain accurate Peach—Koehler forces, the mesh size must be smaller than one-third of the distance separating two dislocation cores. The other method is to calculate the driving force directly from the definition of Peach—Koehler force:

$$\mathbf{F}^{PK}(x_\beta) = \xi(x_\beta) \times (\tilde{\sigma}(x_\beta) \cdot \mathbf{b}_\beta) \tag{10.8}$$

where $\xi(x_\beta)$ and $\mathbf{b}_\beta$ are the direction vector and Burgers vector of dislocation β respectively. $\tilde{\sigma}(x_\beta)$ is the total local stress:

$$\tilde{\sigma}(x_\beta) = \sigma\left(\mathbf{u}^{FE}(x_\beta)\right) + \sum_{\alpha \in A_\beta} \sigma(\mathbf{u}_\alpha^D(x_\beta)) \tag{10.9}$$

The first term on the right-hand side of Eq. (10.9) is the stress from the standard part of the displacement approximation and the second term is from the interaction of neighboring dislocations.

Figure 10.11 shows the X-FEM simulation results for dislocation loop evolution. Compared with the discrete dislocation method in section 10.1.1, it is more difficult to deal with the short-range interaction of quantities of dis-locations, and the topology of dislocations like annihilation, multiplication,

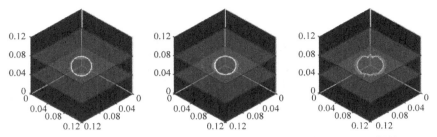

FIGURE 10.11   X-FEM simulation of dislocation loop evolution (Gracie et al., 2008).

etc. Besides this, the choice of an appropriate dislocation core enrichment function in Eq. (10.7) is also a challenge in the X-FEM modeling of dislocations.

10.2 APPLICATION OF MULTI-SCALE SIMULATION

Simulations with atomistic resolution of dislocation cores and crack fronts are critical to a more fundamental understanding of the physics of plasticity and failure. However, even the treatment of submicron cracks and dislocation loops by atomistic methods is generally not feasible today because of the large number of atoms required (Gracie and Belytschko, 2009). To circumvent this problem, many multi-scale computational models have been proposed to simulate defects like cracks, dislocation, etc. Recently, Gracie and Belytschko (2009) successfully applied X-FEM in the atomistic/continuum model to simulate dislocations and cracks. In this method, the crack front and dislocation core are in the atomistic region, and use a continuum model in the rest of the domain. The crack surfaces and dislocation glide planes far from the crack tip and dislocation core are modeled by X-FEM (Gracie and Belytschko, 2009). Figure 10.12 is a schematic of this coupling method. The crack tip and dislocation cores are modeled atomistically. The portion of the crack behind the crack tip and the slip between the crack tip and the dislocation core are modeled by X-FEM. Around each fully atomistic subdomain is a bridging domain where both an atomistic and a continuum model exist. In the bridging subdomains, compatibility between the atomistic and continuum models is enforced. Note that the crack and the glide plane are completely independent of the FEM mesh; the elements are larger than the lattice spacing and the FEM mesh does not conform to the lattice (Gracie and Belytschko, 2009). These advantages of the coupling method improve the computational ability of the current multi-scale method.

Recently, Moseley et al. (2012) further extended this new framework so that as cracks and dislocations propagate, the regions ahead of the crack tips and cores will be converted from continuum to atomistic and the discontinuities in the atomistic domain behind the tips and cores will be coarse-grained.

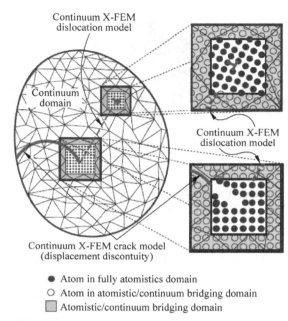

FIGURE 10.12   Schematic of a coupled X-FEM and bridging domain method continuum/atomistic model of an edge dislocation emanating from a crack tip (Gracie and Belytschko, 2009).

Because atomic degrees of freedom are maintained only where needed for each time step, the solution retains the advantages of multi-scale modeling, with a reduced computational cost compared with other multi-scale methods. A crack propagating in a zigzag direction in a sheet of single-layer graphene is simulated using this method, as shown in Figure 10.13.

10.3 MODELING OF DEFORMATION LOCALIZATION

At high strain rates, the dislocation sources in a crystal tend to be localized in one active slip system, and the presence of active dislocation sources in rapidly deforming crystalline material can serve as the focus of localized plastic flow and associated energy concentrations that in turn determine the mechanical response of a material. Most of these localized regions often manifest themselves in the form of deformation bands (Zbib and Diaz de la Rubia, 2002), where there are stress and strain concentration regions in the band because of the heterogeneity of plastic deformation. They are often the locations where further failure occurs, such as the formation of cracks and growth of voids (Liu et al., 2008). As illustrated in Figure 10.14, the loading rate effects on the yield stress and the deformation patterning of single-crystal copper are investigated in the tension simulation. The simulation setup consists of a block with

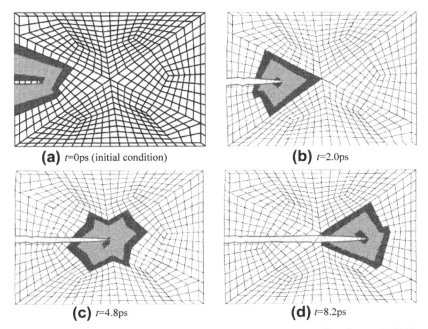

(a) t=0ps (initial condition)

(b) t=2.0ps

(c) t=4.8ps

(d) t=8.2ps

FIGURE 10.13   Modeling of zigzag crack propagation in graphene (Moseley et al., 2012).

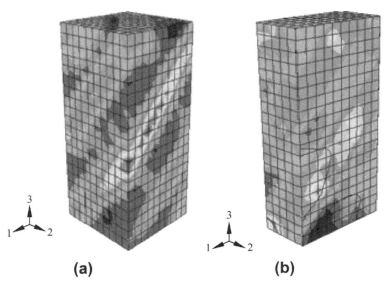

(a)

(b)

FIGURE 10.14   Distribution of strain ε_{33} and Tresca stress in the crystal after yield under strain rate 10^3 s^{-1} (Liu et al., 2008).

dimensions 5 μm × 5 μm 10 μm, which is divided into a finite element mesh using eight-node brick elements. The bottom surface of the sample is fixed, and the upper surface is moved at a controlled rate so that the strain rate, ranging from 10^2 to 10^5 s^{-1}, is made constant (the displacement of the upper surface increases with time). All the other surfaces are assumed to be free.

From the simulation results, it can be observed that the deformation is mostly localized in the bands along the most active slip plane ($1\bar{1}1$) (as shown in Figure 10.14(a)), and with the strain rate increasing, the width of the band also increases from 2 μm to several tens of micrometers, which is restricted by the size of the simulated crystal. In addition, Trasca stress distributions are plotted in Figure 10.14(b), which is relatively higher in the band compared with those in the neighboring regions, and that will induce the materials in the bands to lose the ability to transfer the shear stress, leading to shear instability of the material. So the shear band must be given more attention for crystalline material under high-strain-rate deformation.

The thickness of the deformation band in material usually ranges several tens to hundreds of micrometers, which is far smaller than the size of macro structures. So a very dense mesh is needed in the deformation band to reach the resolution of band width, which is a challenge in FEM simulation (Wright, 2002).

The shear band is a familiar type of deformation band. Inside the band, the tangential component of the displacement is discontinuous. Areias and Belytschko (2007) tried to model the shear band by X-FEM and the displacement field in the elements crossed by the band is expressed as

$$\mathbf{u}^h(\mathbf{x}) = \sum_{I \in S} N_I(\mathbf{x})\mathbf{u}_I + \sum_{J \in S_h} N_J(\mathbf{x})(\mathscr{H}(f(\mathbf{x})) - \mathscr{H}(f(\mathbf{x}_J)))\mathbf{t}\alpha_J(t) \qquad (10.10)$$

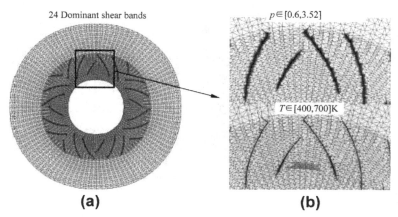

(a)   **(b)**

FIGURE 10.15 Shear band distribution for the thin inner copper cylinder (a), and temperature and effective plastic strains at the Gauss points (b) (Areias and Belytschko, 2007).

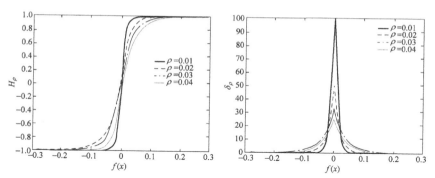

FIGURE 10.16 One-dimensional spatial representation of function $H(f(\mathbf{x}))$ and its derivative (Benvenuti et al., 2008).

where $\mathbf{t}$ is the unit vector in the tangential direction and $\alpha_J(t)$ is the additional degree of freedom. In contrast to crack problems, the shear band is thick and a high strain gradient exists along the thickness direction. In the study of Areias and Belytschko (2007), a linear regularized Heaviside function was constructed to deal with this situation:

$$\mathscr{H}(f(\mathbf{x})) = \begin{cases} 1 & f(\mathbf{x}) > L \\ f(\mathbf{x})/L & |f(\mathbf{x})| \leq L \\ -1 & f(\mathbf{x}) < -L \end{cases} \tag{10.11}$$

where $f(\mathbf{x})$ is the shortest distance between material point $\mathbf{x}$ and the mid-surface of the deformation band, and L denotes the half-thickness of the deformation band. Figure 10.15 shows the X-FEM modeling results of the shear band distribution for the thin inner copper cylinder under shock loading.

Benvenuti et al. (2008) also proposed a nonlinear regularized Heaviside function as the enrichment function:

$$\mathscr{H}(f(\mathbf{x})) = \begin{cases} 1 & f(\mathbf{x}) \rangle \rho \\ \mathrm{sign}(f(\mathbf{x}))(1 - e^{-|f(\mathbf{x})|/\rho}) & |f(\mathbf{x})| \leq \rho \\ 0 & f(\mathbf{x}) < 0 \end{cases} \tag{10.12}$$

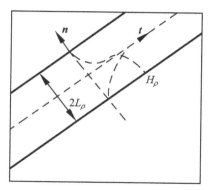

FIGURE 10.17 Function $H(f(\mathbf{x}))$ over the regularization layer of width $2L_\rho$ as a function of the signed distance from the discontinuity line in a two-dimensional quadrilateral domain (Benvenuti et al., 2008).

where ρ is a regularized parameter; the function $\mathscr{H}(f(\mathbf{x}))$ and its derivative for the 1D and 2D cases are plotted in Figures 10.16 and 10.17 respectively.

10.4 SUMMARY

In this chapter, according to the authors' research, new applications and challenges of X-FEM in micro- and nanomechanics, multi-scale computation, and deformation localization are briefly introduced. In addition, X-FEM also has broad potential in contact mechanics, interface mechanics, damage mechanics, etc.

We can conclude from the preceding chapters that X-FEM is prospering and has produced fruitful results in many fields of mechanics. This new method greatly enhances the ability of the conventional finite element method, and more potential applications of X-FEM await.

Westergaard Stress Function Method

One of the analytical methods for finding the stress intensity factor was introduced in section 2.2, the Westergaard stress function method. It is one of the simplest methods for dealing with stress intensity factors. Before the main content is introduced, the essential concepts of elastic mechanics and complex variable functions are briefly reviewed and some practical formulae are provided.

A.1. PLANE PROBLEM AND ANTIPLANE SHEAR PROBLEM IN LINEAR ELASTIC MECHANICS

A.1.1. Definition of Plane Problem

As shown in Figure A.1, the z-axis is the thickness direction of a plate. If the plate is very thick, it is not easily deformed along the z direction, so $\varepsilon_z = \gamma_{xz} = \gamma_{yz} = 0$, which is the plane strain condition. Now, the only strains present are ε_x, ε_y, and γ_{xy}, and the only stresses are σ_x, σ_y, τ_{xy}, and σ_z. Therefore, the relation of stress and strain is given by

$$\varepsilon_z = \frac{1}{E}\left[\sigma_z - \nu(\sigma_x + \sigma_y)\right] = 0 \qquad (A.1)$$

and we may obtain:

$$\sigma_z = \nu(\sigma_x + \sigma_y) \qquad (A.2)$$

where E is elastic modulus and ν is Poisson's ratio.

If a plate is very thin and there is no loading along the z direction, the strain is not restricted along the z direction and we have $\sigma_z \cong \tau_{xz} \cong \tau_{yz} \cong 0$, which is

FIGURE A.1 Coordinate system of a plate.

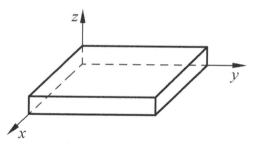

generally the plane stress situation. If the equal sign holds, we have the plane stress condition.

For plane strain and plane stress, the displacement components are u and v corresponding to the x and y directions, without considering the displacement component w corresponding to the z direction. All displacement, strain, and stress components are only functions of x and y coordinates; thus, plane strain and plane stress are two-dimensional problems. They have the same equilibrium equations and geometrical conditions, except for some differences in coefficients in the physical relations. All of these are called plane problems.

If there are only two shear strain components γ_{xz} and γ_{yz} in an elastic body, the stress components have only τ_{xz} and τ_{yz} available, which are functions of x and y. The remaining stress components are zero; this is called the antiplane shear problem. Now, the displacement component w only exists along the z direction, which is also a function of x and y.

A.1.2. Essential Equations for the Plane Problem

(1) Plane Problem

The equilibrium equations are

$$\frac{\partial \sigma_x}{\partial x} + \frac{\partial \tau_{xy}}{\partial y} = 0, \quad \frac{\partial \tau_{xy}}{\partial x} + \frac{\partial \sigma_y}{\partial y} = 0 \tag{A.3}$$

The relations between strain and stress are

$$\varepsilon_x = \frac{\partial u}{\partial x}, \quad \varepsilon_y = \frac{\partial v}{\partial y}, \quad \gamma_{xy} = \frac{\partial u}{\partial y} + \frac{\partial v}{\partial x} \tag{A.4}$$

and the constitution equations are

$$\varepsilon_x = \frac{1}{E_1}\left(\sigma_x - \nu_1 \sigma_y\right), \quad \varepsilon_y = \frac{1}{E_1}\left(\sigma_y - \nu_1 \sigma_x\right), \quad \gamma_{xy} = \frac{\tau_{xy}}{\mu} \tag{A.5}$$

where μ is shear modulus, which is given as follows:

$$E_1 = \begin{cases} \dfrac{E}{1-\nu^2} & \text{Plane strain} \\[2mm] E & \text{Plane stress} \end{cases}$$

$$\nu_1 = \begin{cases} \dfrac{\nu}{1-\nu} & \text{Plane strain} \\[2mm] \nu & \text{Plane stress} \end{cases}$$

Stress functions are defined as

$$\sigma_x = \frac{\partial^2 \Psi}{\partial y^2}, \quad \sigma_y = \frac{\partial^2 \Psi}{\partial x^2}, \quad \tau_{xy} = -\frac{\partial^2 \Psi}{\partial x \partial y} \tag{A.6}$$

By substituting Eq. (A.6) into equilibrium equation (A.3), it can be satisfied automatically.

The compatibility equation is

$$\frac{\partial^2 \varepsilon_x}{\partial y^2} + \frac{\partial^2 \varepsilon_y}{\partial x^2} - \frac{\partial^2 \gamma_{xy}}{\partial x \partial y} = 0 \tag{A.7}$$

By substituting Eqs. (A.5) and (A.6) into compatibility equation (A.7), the double-compatibility equation is obtained as

$$\nabla^4 \Psi = \nabla^2 (\nabla^2 \Psi) = 0 \tag{A.8}$$

As polar coordinates are used, the relations of stress function Ψ and stress components become

$$\sigma_r = \frac{1}{r}\frac{\partial \Psi}{\partial r} + \frac{1}{r^2}\frac{\partial^2 \Psi}{\partial \theta^2}, \quad \sigma_\theta = \frac{\partial^2 \Psi}{\partial r^2}, \quad \tau_{r\theta} = -\frac{\partial}{\partial r}\left(\frac{1}{r}\frac{\partial \Psi}{\partial \theta}\right) \tag{A.9}$$

Now, the double-compatibility equation (A.8) still holds. However, the Laplace operator can be expressed in polar coordinates, such that

$$\nabla^2 = \frac{\partial^2}{\partial r^2} + \frac{1}{r}\frac{\partial}{\partial r} + \frac{1}{r^2}\frac{\partial^2}{\partial \theta^2} \tag{A.10}$$

(2) Antiplane Shear Problem

Another special situation is where there is only a displacement component w along the z direction, and w is only a function of x and y, which is an antiplane shear problem. Now, the existing strains have only relations among γ_{xz} and γ_{yz} with displacement w, such that

$$\gamma_{xz} = \frac{\partial w}{\partial x}, \quad \gamma_{yz} = \frac{\partial w}{\partial y} \tag{A.11}$$

The relations between stress and strain are

$$\tau_{xz} = \mu \gamma_{xz}, \quad \tau_{yz} = \mu \gamma_{yz} \tag{A.12}$$

Substituting Eqs. (A.11) and (A.12) into the equilibrium equation, we get

$$\frac{\partial \tau_{xz}}{\partial x} + \frac{\partial \tau_{yz}}{\partial y} = 0 \tag{A.13}$$

and the compatibility equation is obtained as

$$\nabla^2 w = 0 \tag{A.14}$$

As polar coordinates are used, the compatibility equation (A.14) still holds. However, the strain components are γ_{rz} and $\gamma_{\theta z}$ only. The relations between strain components and displacement are given by

$$\tau_{rz} = \mu\gamma_{rz} = \mu\frac{\partial w}{\partial r}, \quad \tau_{\theta z} = \mu\gamma_{\theta z} = \frac{\mu}{r}\frac{\partial w}{\partial \theta} \tag{A.15}$$

The double-compatibility equation (A.8) and compatibility equation (A.14) are the governing equations of the plane and antiplane shear problems respectively. The stress, strain, and displacement components are solved so long as we obtain the solutions of these governing equations and satisfy the boundary conditions. However, this solution process is actually very complex and difficult; consequently, only limited problems can be solved.

A.1.3. Stress Transformation Relation

Index notation can be used for the general form of stress transformation:

$$\sigma'_{ij} = l_{im}l_{jn}\sigma_{mn} \tag{A.16}$$

where σ'_{ij} and σ_{mn} represent the stress components corresponding to different coordinate systems. On the right-hand side of Eq. (A.16), indices repeated twice in a term are summed to conform with the rules of Einstein notation. As shown in Figure A.2, l_{ij} is the cosine of two orientation angles between i and j, which is called the orientation cosine. When the right-angle coordinate is transformed to a polar coordinate system, we have the orientation cosines below:

$$l_{rx} = \cos\theta, \quad l_{ry} = \sin\theta, \quad l_{\theta x} = -\sin\theta, \quad l_{\theta y} = \cos\theta$$

Therefore, the stress component relations between polar coordinate and right-angle coordinate are

$$\left.\begin{array}{l} \sigma_r = \sigma_x \cos^2\theta + \sigma_y \sin^2\theta + 2\tau_{xy}\cos\theta\sin\theta \\ \sigma_\theta = \sigma_x \sin^2\theta + \sigma_y \cos^2\theta - 2\tau_{xy}\cos\theta\sin\theta \\ \tau_{r\theta} = (\sigma_y - \sigma_x)\cos\theta\sin\theta + \tau_{xy}(\cos^2\theta - \sin^2\theta) \end{array}\right\} \tag{A.17}$$

FIGURE A.2 Coordinate transformation.

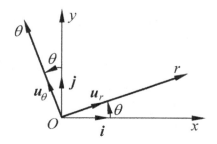

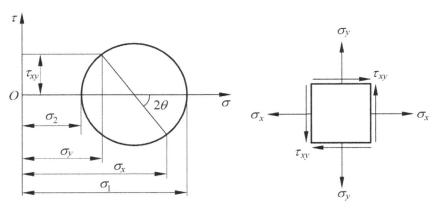

FIGURE A.3   Mohr's circle.

The stress transformation of the antiplane shear problem can also be achieved using Eq. (A.16).

A.1.4. Principal Stresses

Mohr's circle is used, as shown in Figure A.3, to solve principal stresses σ_1 and σ_2 respectively:

$$\frac{\sigma_1}{\sigma_2} = \frac{\sigma_x + \sigma_y}{2} \pm \sqrt{\left(\frac{\sigma_x - \sigma_y}{2}\right)^2 + \tau_{xy}^2} \qquad (A.18)$$

A.2. COMPLEX VARIABLE STRESS FUNCTION

A.2.1. Essential Definition of Complex Variable Function

A complex variable is defined as

$$z = x + iy \qquad (A.19)$$

Its conjugate complex variable is

$$\bar{z} = x - iy \qquad (A.20)$$

A real function is the function of variables x and y. By utilizing Eqs. (A.19) and (A.20) to change variables x and y to z and $\bar{z}$, this function still has a real value when it is written in a new functional form. Therefore, the Airy stress function $\Psi(x,y)$ of the double-compatibility equation can be written as

$$\Psi(x, y) = \Psi(x(z, \bar{z}), y(z, \bar{z})) = \Phi(z, \bar{z}) \qquad (A.21)$$

If there is an antiplane shear problem, the only existing displacement component w can be modified as

$$w(x, y) = W(z, \bar{z}) \tag{A.22}$$

Although w and W have the same value, the functional form is different. In order to express complex variables conveniently, the symbol W is replaced by w in the following discussions.

A.2.2. Plane Problem

Example A.1. Try to prove that the double-compatibility equation (A.8) can be rewritten as

$$\frac{\partial^4 \mathbf{\Phi}(z, \bar{z})}{\partial z^2 \partial \bar{z}^2} = 0 \tag{A.23}$$

Solution. By solving Eqs. (A.19) and (A.20), we obtain

$$\frac{\partial}{\partial x} = \frac{\partial}{\partial z}\frac{\partial z}{\partial x} + \frac{\partial}{\partial \bar{z}}\frac{\partial \bar{z}}{\partial x} = \frac{\partial}{\partial z} + \frac{\partial}{\partial \bar{z}} \tag{a}$$

$$\frac{\partial}{\partial y} = \frac{\partial}{\partial z}\frac{\partial z}{\partial y} + \frac{\partial}{\partial \bar{z}}\frac{\partial \bar{z}}{\partial y} = i\left(\frac{\partial}{\partial z} - \frac{\partial}{\partial \bar{z}}\right) \tag{b}$$

$$\frac{\partial^2}{\partial x^2} = \left(\frac{\partial}{\partial z} + \frac{\partial}{\partial \bar{z}}\right)^2 = \frac{\partial^2}{\partial z^2} + 2\frac{\partial^2}{\partial z \partial \bar{z}} + \frac{\partial^2}{\partial \bar{z}^2} \tag{c}$$

$$\frac{\partial^2}{\partial y^2} = -\left(\frac{\partial}{\partial z} - \frac{\partial}{\partial \bar{z}}\right)^2 = -\left(\frac{\partial^2}{\partial z^2} - 2\frac{\partial^2}{\partial z \partial \bar{z}} + \frac{\partial^2}{\partial \bar{z}^2}\right) \tag{d}$$

Summing Equations (c) and (d) and then taking squares gives the equation:

$$\nabla^4 = \left(\frac{\partial^2}{\partial x^2} + \frac{\partial^2}{\partial y^2}\right)^2 = \left(4\frac{\partial^2}{\partial z \partial \bar{z}}\right)^2 = 16\frac{\partial^4}{\partial z^2 \partial \bar{z}^2} \tag{e}$$

such that

$$\nabla^4 \mathbf{\Psi} = 16\frac{\partial^4 \mathbf{\Phi}}{\partial z^2 \partial \bar{z}^2} = 0$$

So, the double-compatibility equation can be rewritten to become Eq. (A.23) and the proof is complete.

Four times integrations are carried out for Eq. (A,23), and the general solution of this equation can be achieved:

$$\mathbf{\Phi}(z, \bar{z}) = f_1(z) + f_2(\bar{z}) + z f_3(\bar{z}) + \bar{z} f_4(z) \tag{A.24}$$

Because the function Φ has a real value, the following relations between the functions are found:

$$\overline{f_1(z)} = f_2(\overline{z}), \quad f_3(\overline{z}) = \overline{f_4(z)}$$

Thus, in Eq. (A.24), there are only two independent complex variable functions on the right-hand side of the equals sign. The general solution can be rewritten as

$$\Phi(z, \overline{z}) = f_1(z) + \overline{f_1(z)} + z\overline{f_4(z)} + \overline{z}f_4(z) \tag{A.25}$$

To express this more conveniently, we suppose $f_1(z) = \phi(z)/2$ and $f_4(z) = \psi(z)/2$, and the above equation becomes

$$\Phi(z, \overline{z}) = \frac{1}{2}\left[\phi(z) + \overline{\phi(z)} + \overline{z}\psi(z) + z\overline{\psi(z)}\right]$$

or

$$\Phi(z, \overline{z}) = \mathrm{Re}[\phi(z) + \overline{z}\psi(z)] \tag{A.26}$$

where Re indicates the real part of the function and Im indicates the imaginary part. Using Eq. (A.6) as well as Eqs. (c) and (d) in Example A.1, we obtain

$$\sigma_x + \sigma_y = \left(\frac{\partial^2}{\partial x^2} + \frac{\partial^2}{\partial y^2}\right)\Psi = 4\frac{\partial^2 \Phi}{\partial z \partial \overline{z}}$$

In the same way, the following equation is found:

$$\sigma_y - \sigma_x + 2i\tau_{xy} = \left(\frac{\partial^2}{\partial x^2} - \frac{\partial^2}{\partial y^2} - 2i\frac{\partial^2}{\partial x \partial y}\right)\Psi = 4\frac{\partial^2 \Phi}{\partial z^2}$$

By substituting Eq. (A.26) into the above two equations and then solving their derivatives, the following results are obtained:

$$\sigma_x + \sigma_y = 2\left[\psi'(z) + \overline{\psi'(z)}\right] \tag{A.27}$$

$$\sigma_y - \sigma_x + 2i\tau_{xy} = 2[\phi''(z) + \overline{z}\psi''(z)] \tag{A.28}$$

The same method is used to give the displacement expression of the complex variable function. Since the deduction process is complex, it is not shown here and is left for the exercises. The displacement expression is given by

$$2\mu(u + iv) = \kappa\psi(z) - z\overline{\psi'(z)} - \overline{\phi'(z)} \tag{A.29}$$

For a plane strain problem, $\kappa = 3 - 4\nu$; for a plane stress problem, $\kappa = (3 - \nu)/(1 + \nu)$.

A.2.3. Antiplane Shear Problem

Following the method of the plane problem, we can obtain the displacement and stress component expressions of complex variable functions for an antiplane shear problem:

$$\omega = \text{Re}[f(z)] \tag{A.30}$$

$$\tau_{xz} - i\tau_{yz} = \mu f'(z) \tag{A.31}$$

By again using the complex variable stress function, the original problem of solving a double-compatibility equation or compatibility equation is resolved into a problem of finding the complex variable stress function that satisfies the boundary conditions. For mode I and mode II fracture problems, Eqs. (A.27)–(A.29) can be simplified to a single expression for the complex variable function, which is called the Westergaard stress function.

A.3. WESTERGAARD STRESS FUNCTION

A.3.1. Mode I Fracture

Along the x-axis, if $\tau_{xy} = 0$ and $\sigma_x = \sigma_y$, by solving Eq. (A.28) the following equation can be obtained:

$$\left[\sigma_y - \sigma_x + 2i\tau_{xy}\right]_{y=0} = 2[\phi''(x) + x\psi''(x)] = 0$$

By taking $\phi''(z) = -z\psi''(z)$, the above equation is satisfied. Therefore, Eq. (A.28) can be expressed by a second-order derivative of complex variable function $\psi(z)$, such that

$$\sigma_y - \sigma_x + 2i\tau_{xy} = -4iy\psi''(z) \tag{A.32}$$

Combining Eqs. (A.27) and (A.32), the solution is

$$\begin{cases} \sigma_x = 2\{\text{Re}[\psi'(z)] - y\,\text{Im}[\psi''(z)]\} \\ \sigma_y = 2\{\text{Re}[\psi'(z)] + y\,\text{Im}[\psi''(z)]\} \\ \tau_{xy} = -2y\text{Re}[\psi''(z)] \end{cases}$$

A complex variable function is defined as

$$Z_I(z) = 2\psi'(z)$$

Substituting this function into the above solution, we get

$$\begin{cases} \sigma_x = \text{Re}[Z_I] - y\,\text{Im}[Z'_I] \\ \sigma_y = \text{Re}[Z_I] + y\,\text{Im}[Z'_I] \\ \tau_{xy} = -y\text{Re}[Z'_I] \end{cases} \tag{A.33}$$

where

$$Z'_I(z) = \frac{dZ_I(z)}{dz} \tag{A.34}$$

If we take

$$Z_I(z) = \frac{d\tilde{Z}_I(z)}{dz} \tag{A.35}$$

then the displacement expression (A.29) becomes

$$\begin{cases} 2\mu u = \dfrac{\kappa - 1}{2} \mathrm{Re}\left[\tilde{Z}_I\right] - y\,\mathrm{Im}[Z_I] \\[3mm] 2\mu v = \dfrac{\kappa + 1}{2} \mathrm{Im}\left[\tilde{Z}_I\right] - y\,\mathrm{Re}[Z_I] \end{cases} \tag{A.36}$$

For a mode I crack, if the crack lies on the x-axis, we have $\tau_{xy}\big|_{y=0} = 0$, while σ_x may not always equal σ_y. However, a normal stress may always be superposed on it along the x-axis, which results in the relation $\sigma_x = \sigma_y$ along the x-axis. Because stress singularity exists at a crack tip location, it does not affect the stress field and displacement field at the crack tip when a finite normal stress is superposed on it.

A.3.2. Mode II Fracture

For a mode II crack, it is now restricted by $\sigma_y\big|_{y=0} = 0$. By following the method for the mode I crack and taking

$$\psi'(z) = -\frac{i}{2} Z_{II}(z) \tag{A.37}$$

we get

$$\begin{cases} \sigma_x = 2\mathrm{Im}\,Z_{II} + y\mathrm{Re}\,Z'_{II} \\ \sigma_y = -y\mathrm{Re}\,Z'_{II} \\ \tau_{xy} = \mathrm{Re}\,Z_{II} - y\mathrm{Im}\,Z'_{II} \end{cases} \tag{A.38}$$

$$\begin{cases} 2\mu u = \dfrac{\kappa + 1}{2} \mathrm{Im}\,\tilde{Z}_{II} + y\mathrm{Re}\,Z_{II} \\[3mm] 2\mu v = -\dfrac{\kappa - 1}{2} \mathrm{Re}\,\tilde{Z}_{II} - y\mathrm{Im}\,Z_{II} \end{cases} \tag{A.39}$$

in which the meaning of the sign is the same as for the mode I crack, which is only abandoned in the middle brackets of the real and imaginary parts.

The above solutions were first obtained by the famous mechanics scientist Westergaard. Therefore, Z_I and Z_{II} are called Westergaard stress functions. It is difficult to get the Westergaard stress function to satisfy the boundary condition; that is beyond the scope of this book. Thus, the Westergaard stress function is provided in the following discussion, and it is verified to check whether the boundary condition of cracked problem is satisfied. Finally, Eq. (A.33) or (A.38) is used to solve the stress field at the crack tip and obtain the stress intensity factor.

A.4. ESSENTIAL FRACTURE PROBLEMS

It is relatively easy to solve a cracked infinite plate problem using the Westergaard stress function method. The following examples belong to this kind of problem.

Example A.2. An infinite plate including a center crack penetrating into its thickness is considered, which is subjected to double-direction uniform tension stresses far away, as shown in Figure A.4. Try to verify the function

$$Z_I(z) = \frac{\sigma_z}{\sqrt{z^2 - a^2}} \tag{A.40}$$

which is the Westergaard stress function, satisfying the boundary condition, and find the stress field and displacement field at the crack tip location.

Solution. As illustrated in Figure A.4, if there is no tension stress in the x direction, we have a Griffith fracture problem. The stress function given by Eq. (A.40) is a function in the right-angle coordinate system, which takes the crack center as the original point. The uniform stress is σ and the semi-length of the crack is a. From Eq. (A.40), we get

$$Z'_I(z) = \frac{-\sigma a^2}{(z^2 - a^2)^{3/2}} \tag{a}$$

$$\tilde{Z}_I(z) = \sigma \sqrt{z^2 - a^2} \tag{b}$$

FIGURE A.4 An infinite plate including a center crack penetrating into its thickness and subjected to double-direction uniform tension stresses.

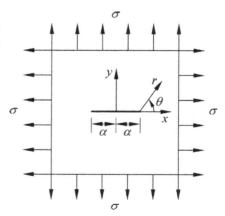

By using the equations given in Eq. (A.33), we need to check whether they satisfy the following boundary conditions and special conditions:

1. When $|x| \to \infty$, $(\sigma_x)_\infty \to \sigma$
2. When $|y| \to \infty$, $(\sigma_y)_\infty \to \sigma$
3. When $y = 0$, $\tau_{xy} = 0$, and $\sigma_x = \sigma_y$
4. When $y = 0$ and $|x| \langle a$, $\sigma_y = \tau_{xy} = 0$, the crack surface is stress free.

It is necessary to satisfy condition 3 if Eq. (A.33) is to be used, which is the condition of geometric symmetry and loading symmetry. It is worth noting that σ_y is equal to zero on the crack surface, while σ_x is not defined on the crack surface. Thus, condition 3 may still be satisfied.

By using Eqs. (A.33), (A.40), and (a) in Example A.2, it is easy to verify boundary conditions 1 and 2 respectively.

When $y = 0$, then $\tau_{xy} = 0$ is found. Thus, finally we verify $\sigma_y = 0$ on the crack surfaces ($|x| \langle a$). The verification is given below:

$$\sigma_y(x,\ 0) = \mathrm{Re}[Z_{\mathrm{I}}]_{y=0} = \mathrm{Re}\left[\frac{\sigma\,x}{\sqrt{x^2 - a^2}}\right]$$

$$= \mathrm{Re}[\text{pureimaginary number function}] = 0$$

Therefore, the solution obtained by Eq. (A.40) and stress expression (A.33), displacement expression (A.36) is the accurate solution. Since the crack is symmetrical about the y-axis, it is sufficient to consider the stress field and displacement field at only one crack tip zone. Here, a crack tip of $z = a$ is considered. We have

$$z = a + re^{i\theta} \tag{c}$$

Here, (r, θ) are polar coordinates of the chosen crack tip $z = a$ as the original point. At the crack tip zone, where $r << a$, it takes the first term,

$$Z_{\mathrm{I}} = \frac{\sigma(a + re^{i\theta})}{\sqrt{(a + re^{i\theta})^2 - a^2}} \approx \frac{\sigma\sqrt{a}}{\sqrt{2r}}e^{-\frac{i}{2}\theta} = \frac{\sigma\sqrt{a}}{\sqrt{2r}}\left(\cos\frac{\theta}{2} - i\sin\frac{\theta}{2}\right) \tag{d}$$

$$Z'_{\mathrm{I}} = -\frac{\sigma a^2}{\left[(a + re^{i\theta})^2 - a^2\right]^{3/2}} \approx -\frac{\sigma\sqrt{a}}{(2r)^{3/2}}e^{-\frac{i3}{2}\theta} = -\frac{\sigma\sqrt{a}}{(2r)^{3/2}}\left(\cos\frac{3\theta}{2} - i\sin\frac{3\theta}{2}\right) \tag{e}$$

$$\tilde{Z}_{\mathrm{I}} = \sigma\sqrt{(a + re^{i\theta})^2 - a^2} = \sigma\sqrt{2ar}\left(\cos\frac{\theta}{2} + i\sin\frac{\theta}{2}\right) \tag{f}$$

According to the equations in Eq. (A.33) and $y = r\sin\theta$, the stresses at the crack tip zone can be found:

$$\sigma_x = \frac{\sigma\sqrt{a}}{\sqrt{2r}}\cos\frac{\theta}{2}\left[1 - \sin\frac{\theta}{2}\sin\frac{3\theta}{2}\right] \tag{g}$$

$$\sigma_y = \frac{\sigma\sqrt{a}}{\sqrt{2r}}\cos\frac{\theta}{2}\left[1 + \sin\frac{\theta}{2}\sin\frac{3\theta}{2}\right] \tag{h}$$

$$\tau_{xy} = \frac{\sigma\sqrt{a}}{\sqrt{2r}}\sin\frac{\theta}{2}\cos\frac{\theta}{2}\cos\frac{3\theta}{2} \tag{i}$$

Based on the equations in Eq. (A.36), the displacements at the crack tip area can be deduced:

$$2\mu u = \sigma\sqrt{a}\left(\frac{r}{2}\right)^{1/2}\left[(\kappa - 1) + 2\sin^2\frac{\theta}{2}\right]\cos\frac{\theta}{2} \tag{j}$$

$$2\mu v = \sigma\sqrt{a}\left(\frac{r}{2}\right)^{1/2}\left[(\kappa + 1) - 2\cos^2\frac{\theta}{2}\right]\sin\frac{\theta}{2} \tag{k}$$

Comparing the three equations (g)–(i) with Eq. (2.6), it can be seen that the stress intensity factor is given by

$$K_I = \sigma\sqrt{\pi a} \tag{A.41}$$

Example A.3. An infinite plate has a center crack as shown in Figure A.5(a). A pair of tension concentration forces F (force/unit thickness) is acting on the crack surface $x = b$ position. Try to prove the function

$$Z_I = \frac{F\sqrt{a^2 - b^2}}{\pi(z - b)\sqrt{z^2 - a^2}} \tag{l}$$

which is the Westergaard stress function, satisfying the boundary condition, and solve the stress field and displacement field at the crack tip location, as well as the stress intensity factor.

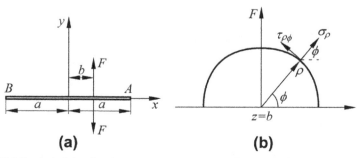

(a) **(b)**

FIGURE A.5 An infinite plate has a center crack subjected to a pair of tension concentration forces. (a) Center crack surface has concentration force. (b) Semi-circular element of concentration force position.

Solution. According to the loading symmetry situation as shown in Figure A.5(a), clearly this is an opening (mode I) crack problem. The boundary condition of the crack surface subjected to a pair of tension concentration forces F at $x = b$ is only verified here. Readers may follow the previous example to deduce the rest by themselves. As shown in Figure A.5(b), we take $z = b$ as an original point and draw a semi-circle with radius ρ. When $\rho \to 0$, the external force and internal force on the whole semi-circle must be kept in equilibrium. Therefore, the equilibrium conditions in the x and y directions are respectively given by

$$\lim_{\rho \to 0} \left[F + \int_0^\pi (\sigma_\rho \sin \phi + \tau_{\rho\phi} \cos \phi) \rho d\phi \right] = 0 \qquad \text{(m)}$$

$$\lim_{\rho \to 0} \left[\int_0^\pi (\sigma_\rho \cos \phi - \tau_{\rho\phi} \sin \phi) \rho d\phi \right] = 0 \qquad \text{(n)}$$

Now, the equations are substituted in Eq. (A.33) given by the Westergaard method and stress function (l) is used to verify whether the conditions (m) and (n) are satisfied. At the location $z = b$, we have

$$z' = \rho e^{i\phi}$$

Then $z = b + z'$, where $\rho << b$ and $\rho << a$. Thus, the solution is obtained as

$$Z_I = \frac{F\sqrt{a^2 - b^2}}{\pi(z - b)\sqrt{z^2 - a^2}} \approx \frac{F\sqrt{a^2 - b^2}}{\pi z' \sqrt{b^2 - a^2}} = \frac{F}{i\pi z'}$$

$$Z'_I = \frac{dZ_I}{dz} = \frac{dZ_I}{dz'} \approx \frac{-F}{i\pi z'^2} = \frac{-Fe^{-2i\phi}}{i\pi \rho^2}$$

The real and imaginary parts of Z_I and Z'_I can be deduced separately as

$$\text{Re } Z_I \approx -\frac{F}{\pi\rho} \sin \phi, \quad \text{Im } Z_I \approx -\frac{F}{\pi\rho} \cos \phi$$

$$\text{Re } Z'_I \approx \frac{F}{\pi\rho^2} \sin 2\phi, \quad \text{Im } Z'_I \approx \frac{F}{\pi\rho^2} \cos 2\phi$$

Thus, the stress field at $z = b$ is

$$\sigma_x = \text{Re } Z_I - y \text{ Im } Z'_I = -\frac{2F \sin \phi \cos^2 \phi}{\pi\rho}$$

$$\sigma_y = \text{Re } Z_I + y \text{ Im } Z'_I = -\frac{2F \sin^3 \phi}{\pi\rho}$$

$$\tau_{xy} = -y \text{ Re } Z'_I = -\frac{2F \sin^2 \phi \cos \phi}{\pi\rho}$$

In the above equations, the approximation sign is replaced by an equal sign. The solutions can be found via stress transformation:

$$\sigma_p = \sigma_x \cos^2 \phi + \sigma_y \sin^2 \phi + 2 \sin \phi \cos \phi \tau_{xy} = -\frac{2P \sin \phi}{\pi \rho} \qquad \text{(o)}$$

$$\tau_{\rho\phi} = -\sigma_x \cos \phi \sin \phi + \sigma_y \cos \phi \sin \phi + \tau_{xy}(\cos^2 \phi - \sin^2 \phi) = 0 \qquad \text{(p)}$$

Substituting Eqs. (o) and (p) into Eqs. (m) and (n), the verification can be fulfilled. For the stress field and displacement field at the crack tip position $z = a$, the reader can follow the steps provided in Example A.2. The difference here lies only in the form of the stress intensity factor. At this moment, the stress intensity factor at point A $(z = a)$ is

$$K_A = \frac{F}{\sqrt{\pi a}} \sqrt{\frac{a+b}{a-b}} \qquad \text{(A.42)}$$

The stress intensity factor at point B can be obtained if b in Eq. (A.42) is replaced by $-b$, such that

$$K_B = \frac{F}{\sqrt{\pi a}} \sqrt{\frac{a-b}{a+b}} \qquad \text{(A.43)}$$

Equations (A.42) and (A.43) are the expressions of stress intensity factor provided in the third row of Table 2.1 in Chapter 2, which are analytically proven here.

The forms of stress field and displacement field are usually fixed at the crack tip location. Their value at some point is fully determined by the stress intensity factor. Once the stress intensity factor is found, it is equivalent to providing the stress field and displacement field at the crack tip location.

EXERCISE A

A.1 If the boundary condition at infinite distance is $\tau_{xy} = \tau$, as shown in Figure A.4, try to prove $Z_{II} = \tau z / \sqrt{z^2 - a^2}$, which is the Westergaard stress function for this problem, and find the stress field, displacement field, and stress intensity factor at the crack tip zone.

A.2 As shown in Figure A.5(a), if the concentration force F is no longer perpendicular to the crack surface, but the force F acting on the upper surface of the crack is along the positive x direction, while another force F acting on the lower surface of the crack is along the negative x direction, try to prove

$$Z_{II} = F\sqrt{a^2 - b^2} / \left[\pi(z - b)\sqrt{z^2 - a^2} \right]$$

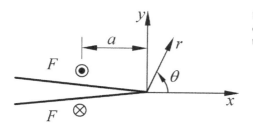

FIGURE A.6 A semi-infinite-length crack subjected to antiplane concentration forces.

which is the Westergaard stress function for this problem, and find the stress field, displacement field, and stress intensity factor at the crack tip zone.

A.3 Try to prove that the crack configuration becomes elliptical for a center crack subjected to double-direction uniform tension, as shown in Figure A.4.

A.4 Consider a semi-infinite-length crack in an infinite plate, as shown in Figure A.6. A pair of concentration forces F, at a distance a from the crack tip and perpendicular to the xy plane, is acting on the upper and lower crack surfaces respectively. Try to prove

$$f'(z) = -\frac{iF\sqrt{a}}{\pi\mu(z+a)\sqrt{z}}$$

which is the Westergaard stress function satisfying the boundary condition, and find the stress field, displacement field, and stress intensity factor at the crack tip zone.

J Integration

The field of fracture mechanics is concerned with the quantitative description of the mechanical state of a deformable body containing a crack or cracks, with a view toward characterizing and measuring the material resistance to crack propagation. The fracture of materials is mainly divided into two branches: brittle fracture and ductile fracture. For example, cast iron, super-steel, stiff aluminum, titanium alloy, organic glass, and epoxy are all brittle materials. However, the vast majority of metallic materials have a middle or lower stiffness, which experiences a large and plastic deformation at the crack tip location before rupture that cannot be ignored. It is no longer appropriate to use linear elastic fracture mechanics. Therefore, it is necessary to adopt an elastic–plastic mechanics method to deal with fracture problems for this kind of ductile material. However, it is difficult and complicated to apply elastic–plastic mechanics to directly obtain stress and strain field at crack tip locations. An alternative to directly solving stress and strain fields at the crack tip location is required, and we need to find a mechanical parameter. This parameter may comprehensively measure the intensity of stress and strain fields at the crack tip location. Based on this parameter, a fracture toughness criterion is established, and finally an experimental scheme is set up to verify the reliability of the theory. The *J* integral described in Chapters 2 and 3 has been developed according in the above-mentioned fashion. The computational method of the *J* integral is given in detail in this Appendix.

B.1 PHYSICAL SIGNIFICANCE OF THE *J* INTEGRAL

A unit thickness plate is shown in Figure B.1. Considering an integral route *C*, which is smooth and has no intersection surrounding the crack tip, the area surrounded is on the left side of the orientation. The integral route starts from a point *F* below the crack and along an anticlockwise direction arrives at a point *F'*, which is just opposite point *F*. A length increment on the integral route is

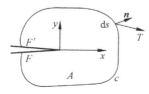

FIGURE B.1   Integral route *C*.

indicated by ds, on which the surface force is acting. The unit vector of external normal is $\boldsymbol{n}$. The surrounded interior area is A.

It has been pointed out by Sanders and Rice that the working rate from the external to the internal part of the route is greater than or equal to the sum of the rate of change of the interior energy stored in A and the unrecovered dissipated energy rate, which can be expressed as

$$\int_C T_i \frac{\mathrm{d}u_i}{\mathrm{d}t} \mathrm{d}s \geq \frac{\mathrm{d}}{\mathrm{d}t} \int_A W_1 \mathrm{d}A + \frac{\mathrm{d}D}{\mathrm{d}t} \tag{B.1}$$

where T_i is a surface force component and is related to the stress as $T_i = \sigma_{ij} n_j$, in which n_j is the projection of $\boldsymbol{n}$ along the x or y direction; u_i is the displacement component, W_1 is interior energy density, and D is dissipated energy.

In the first term of Eq. (B.1), the indices i repeated twice are summed to conform with the rules of Einstein notation. When a "greater than" sign appears, it indicates the crack growth and the kinetic energy variation. If it assumes a quasi-steady state the equals sign holds. It is also assumed that the dissipated energy can only be used to produce a new crack area; then, we have $\mathrm{d}D/\mathrm{d}t = G \cdot \mathrm{d}a/\mathrm{d}t$, where G is a parameter and a is crack length (or semi-length). Thus, Eq. (B.1) becomes

$$\int_C T_i \frac{\mathrm{d}u_i}{\mathrm{d}t} \mathrm{d}s = \frac{\mathrm{d}}{\mathrm{d}t} \int_A W_1 \mathrm{d}A + G \frac{\mathrm{d}a}{\mathrm{d}t} \tag{a}$$

Because it is a quasi-static crack, a may be a variable length, such that

$$\frac{\mathrm{d}}{\mathrm{d}t} = \frac{\partial}{\partial a} \frac{\mathrm{d}a}{\mathrm{d}t} \tag{b}$$

Substituting Eq. (b) into Eq. (a), we obtain

$$\int_C T_i \frac{\partial u_i}{\partial a} \cdot \frac{\mathrm{d}a}{\mathrm{d}t} \mathrm{d}s = \frac{\mathrm{d}a}{\mathrm{d}t} \cdot \frac{\partial}{\partial a} \int_A W_1 \mathrm{d}A + G \frac{\mathrm{d}a}{\mathrm{d}t} \tag{c}$$

At quasi-steady state, da/dt can be regarded as a quantity that is not equal to zero and is independent of coordinates. It can be moved outside the integral sign and rearranged. Therefore, Eq. (c) becomes

$$\int_C T_i \frac{\partial u_i}{\partial a} \mathrm{d}s = \frac{\partial}{\partial a} \int_A W_1 \mathrm{d}A + G \tag{d}$$

or

$$\int_C T_i \frac{\partial u_i}{\partial a} \mathrm{d}s = \int_A \frac{\partial W_1}{\partial a} \mathrm{d}A + G \tag{e}$$

As can be seen in Figure B.2, if the original point is taken throughout at the crack tip, then along with the crack's quasi-static growth, when a is increased, x decreases, so

$$dx = -da \quad \text{or} \quad \frac{\partial}{\partial x} = -\frac{\partial}{\partial a} \tag{f}$$

Shifting the position of each term and using Eq. (f), Eq. (e) then becomes

$$G = \int_A \frac{\partial W_1}{dx} dA - \int_C T_i \frac{\partial u_i}{dx} ds \tag{g}$$

Although integral route C is not closed, the clearance between the upper and lower parts of the crack is regarded as zero, and if we consider the symmetry of the plane crack problem or antiplane crack problem, the interior energy density W_1 is the same at points F and F'. Consequently, the Green theorem of mathematics can be applied here, to give

$$\int_A \left(\frac{\partial Q}{\partial x} - \frac{\partial P}{\partial y} \right) dxdy = \int_C Pdx + Qdy \tag{B.2}$$

Now, $Q = W_1$ and $P = 0$, so Eq. (g) becomes

$$G = \int_C W_1 dy - T_i \frac{\partial u_i}{dx} ds \tag{h}$$

It can be seen that Eq. (B.1) is a form of Irwin–Orowan energy balance equation around the crack tip zone. An essential idea is still the Griffith energy release concept, where energy released at the crack tip forms a new crack area. So the energy release rate G must be related to the surface energy γ_p. If the value on the right-hand side of Eq. (h) is not large enough to form a new crack area, G may be used to measure the tendency of energy release capability. According to the meaning of Eq. (B.1) and the deduction of Sanders, G is the Griffith energy release rate for a linear elastic body. For a linear elastic body containing a crack, if the internal energy is equivalent to the strain energy, then G represents a mechanical parameter to compositely measure the intensity of the stress and strain field on the interior of the route. Equation (h) is first

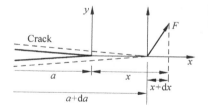

FIGURE B.2   Coordinate system of crack quasi-static growth.

obtained based on the research work of Rice and Sanders; thus, it uses the letter J in the name of James Rice to represent G. For convenience, the symbol W_1 is replaced by W, which is the strain energy density. Therefore, Eq. (h) becomes the Rice integral expression:

$$J = \int_C W \, dy - T_i \frac{\partial u_i}{\partial x} \, ds \tag{B.3}$$

In the following discussion, taking a mode I crack of a plane strain problem as an example, it is proved that the J integral is equal to energy release rate G for a linear elastic body if the integral route C is taken very close to the crack tip. As shown in Figure B.3, the variables are expressed in terms of polar coordinates:

$$ds = r \, d\theta, \quad dy = r \cos \theta \, d\theta$$

So, Eq. (B.3) becomes

$$J = r \int_{-\pi}^{\pi} \left(W \cos \theta - T_i \frac{\partial u_i}{\partial x} \right) d\theta \tag{B.4}$$

If the integral route is taken very close to the crack tip, the radius r is very small. Consequently, the stress and strain field at the crack tip area can be used for mode I cracks in linear elastic fracture mechanics.

For a plane strain problem, the strain energy density becomes

$$W = \frac{k^2}{16\mu r} (3 - 4v - \cos \theta)(1 + \cos \theta)$$

Component forces T_x and T_y of surface force T exist on route C. According to $n_x = \cos \theta$ and $n_y = \cos \theta$ (see Figure B.3) and considering Eq. (2.6), we have

$$T_x = \sigma_x n_x + \tau_{xy} n_y = \frac{k}{\sqrt{2r}} \left[\cos \theta \cos \frac{\theta}{2} - \frac{1}{2} \sin \theta \sin \frac{\theta}{2} \right]$$

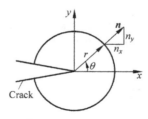

FIGURE B.3 Integral circular route.

$$T_x = \sigma_x n_x + \tau_{xy} n_y = \frac{k}{\sqrt{2r}} \left[\cos \theta \cos \frac{\theta}{2} - \frac{1}{2} \sin \theta \sin \frac{\theta}{2} \right]$$

$$W = \frac{k^2}{16\mu r} (3 - 4\nu - \cos \theta)(1 + \cos \theta),$$

$$T_y = \sigma_y n_y + \tau_{xy} n_x = \frac{k}{\sqrt{2r}} \left[\sin \theta \cos \frac{\theta}{2} - \frac{1}{2} \sin \theta \cos \frac{\theta}{2} \right]$$

By making use of $\partial r / \partial x = \cos \theta$ and $\partial \theta / \partial x = -\sin \theta / r$, as well as Eq. (2.23), the partial derivative of displacement components can be found to be

$$\frac{\partial u}{\partial x} = \frac{\partial u}{\partial r} \cdot \frac{\partial r}{\partial x} + \frac{\partial u}{\partial \theta} \cdot \frac{\partial \theta}{\partial x}$$

$$= \frac{\partial u}{\partial r} \cos \theta + \frac{\partial u}{r \partial \theta} \cdot \sin \theta$$

$$= \left(\frac{1 + \nu}{E} \right) \frac{k}{\sqrt{2r}} \cos \frac{\theta}{2} \left(\frac{3}{2} - 2\nu - \frac{1}{2} \cos \theta - \sin^2 \theta \right)$$

$$\frac{\partial v}{\partial x} = \frac{\partial v}{\partial r} \cdot \frac{\partial r}{\partial x} + \frac{\partial v}{\partial \theta} \cdot \frac{\partial \theta}{\partial x}$$

$$= \frac{\partial v}{\partial r} \cos \theta - \frac{\partial v}{r \partial \theta} \cdot \sin \theta$$

$$= \left(\frac{1 + \nu}{E} \right) \frac{k}{\sqrt{2r}} \sin \frac{\theta}{2} \left(-\frac{3}{2} + 2\nu + \frac{1}{2} \cos \theta - \sin^2 \theta \right)$$

By integrating the first term of Eq. (B.4), we obtain

$$I_1 = r \int_{-\pi}^{\pi} W \cos \theta d\theta = \frac{\pi(1 + \nu)(1 - 2\nu)k^2}{4E}$$

and integrating the second term of Eq. (B.4), the result is

$$I_2 = -r \int_{-\pi}^{\pi} T_i \frac{\partial u_i}{dx} d\theta = -\frac{\pi(1 + \nu)\left(-\frac{3}{2} + \nu\right)k^2}{2E}$$

Finally, the *J* integral value is obtained by I_1 plus I_2:

$$J = I_1 + I_2 = \frac{\pi(1 - \nu^2)k^2}{E} = G_I \tag{B.5}$$

Equation (B.5) has proved that the *J* integral is equal to energy release rate *G* at the crack tip area for the plane strain mode I crack problem of a linear elastic body.

The integral route in the above proof takes the crack tip as an original point of the circular route. Is the same calculation result obtained if it selects a different route? In the following section, it is proved that the value of J integration is independent of the selected route. Therefore, no matter what integral route is chosen, the value of the J integral is equal to the energy release rate for plane strain mode I crack problems of linear elastic bodies. This relation also holds for the plane stress problem. For mode II and mode III cracks, as well as complex cracks, similar relations can be also obtained. So we have the equations:

$$J_K = G_K \quad (K = \text{I, II, III}) \tag{B.6}$$

These relations are established based upon the Irwin hypothesis, such that the crack propagates along the original direction; consequently, the relation between G and stress intensity factor is established.

If the plastic zone at the crack tip is too large to ignore, is the calculated value of the J integral still the same as the result for a linear elastic body? It can be imagined that, when the plastic deformation is small (small-scale yielding), the stress and displacement field outside the plastic zone still can be expressed as the linear elastic results although the stress and displacement field in the plastic zone are unknown. Hence, Eq. (B.6) still holds. If the plastic deformation is relatively large, the stress intensity factor no longer expresses the intensity of the stress field at the crack tip, and the stress and displacement fields outside the plastic zone given by linear elastic mechanics are not valid. Hence, Eq. (B.6) no longer holds. Now, the J integral becomes a mechanical parameter to measure the intensity of the stress and strain fields during plastic deformation. Since the J integral can express the situation of elastic and plastic deformation, and avoid directly calculated complicated stress and displacement fields at the crack tip, the J integral is a very useful parameter in fracture mechanics.

B.2 ROUTE INDEPENDENCE OF THE *J* INTEGRAL

In the process of proving $J = G$, as a result of using the circular route close to the crack tip, the value of the J integral is independent of the selected route. So long as an initial point on the route lies in the crack surface, and the route is smooth and has no point of intersection, then the value of the J integral is invariant. In the following discussion, the route independence of the J integral is validated.

Two anticlockwise integral routes C_1 and C_2 are selected arbitrarily, as shown in Figure B.4, and another two integral routes on the crack surface C_3 (from F' to Q') and C_4 (from Q to F) are added. The route is defined by $C = C_1 + C_3 - C_2 + C_4$, which is a closed-loop contour. The area in the closed contour is A. Now, we need to prove

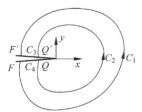

FIGURE B.4   Proof that integral loop taken is independent of linear route.

$$\int_{C_1} W\mathrm{d}y - T_i \frac{\partial u_i}{\partial x}\,\mathrm{d}s = \int_{C_2} W\mathrm{d}y - T_i \frac{\partial u_i}{\partial x}\,\mathrm{d}s \tag{a}$$

Firstly,

$$\int_{C} W\mathrm{d}y - T_i \frac{\partial u_i}{\partial x}\,\mathrm{d}s = 0 \tag{b}$$

By using Green theorem, Eq. (B.2), we obtain

$$\int_{C} W\mathrm{d}y = \int_{A} \frac{\partial W}{\partial x}\,\mathrm{d}x\mathrm{d}y \tag{c}$$

Because the elastic–plastic strain energy density is a function of strain components, no matter if the relation of stress and strain is linear or nonlinear, we always have

$$\partial W / \partial \varepsilon_{ij} = \sigma_{ij} \tag{d}$$

Thus, Eq. (c) may become

$$\int_{C} W\mathrm{d}y = \int_{A} \frac{\partial W}{\partial \varepsilon_{ij}}\frac{\partial \varepsilon_{ij}}{\partial x}\,\mathrm{d}x\mathrm{d}y = \int_{A} \sigma_{ij}\frac{\partial \varepsilon_{ij}}{\partial x}\,\mathrm{d}x\mathrm{d}y \tag{e}$$

As shown in Figure B.5, the relation between the arbitrary integral element ds on the route and outer normal is given by

$$\mathrm{d}y = n_x \mathrm{d}s, \qquad \mathrm{d}x = -n_y \mathrm{d}s \tag{f}$$

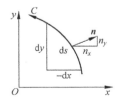

FIGURE B.5   Unit integral element on the route and unit outer normal vector.

Therefore,

$$
\int_C T_i \frac{\partial u_i}{\partial x} ds = \int_C \sigma_{ij} n_j \frac{\partial u_i}{\partial x} ds = \int_C \left[(\sigma_x n_x + \tau_{xy} n_y) \frac{\partial u}{\partial x} + (\sigma_y n_y + \tau_{xy} n_x) \frac{\partial v}{\partial x} \right] ds
$$

$$
= \int_C \left[\left(\sigma_x \frac{\partial u}{\partial x} + \tau_{xy} \frac{\partial v}{\partial x} \right) n_x + \left(\sigma_y \frac{\partial v}{\partial x} + \tau_{xy} \frac{\partial u}{\partial x} \right) n_y \right] ds
$$

$$
= \int_C \left(\sigma_x \frac{\partial u}{\partial x} + \tau_{xy} \frac{\partial v}{\partial x} \right) dy - \left(\sigma_y \frac{\partial v}{\partial x} + \tau_{xy} \frac{\partial u}{\partial x} \right) dx
$$

(g)

Using Green's theorem, the line integral is transformed into a surface integral:

$$
\int_C T_i \frac{\partial u_i}{\partial x} ds = \int_A \left[\frac{\partial}{\partial x} \left(\sigma_x \frac{\partial u}{\partial x} + \tau_{xy} \frac{\partial v}{\partial x} \right) + \frac{\partial}{\partial y} \left(\sigma_y \frac{\partial v}{\partial x} + \tau_{xy} \frac{\partial u}{\partial x} \right) \right] dxdy
$$

$$
= \int_A \left[\left(\frac{\partial \sigma_x}{\partial x} + \frac{\partial \tau_{xy}}{\partial y} \right) \frac{\partial u}{\partial x} + \left(\frac{\partial \tau_{xy}}{\partial x} + \frac{\partial \sigma_y}{\partial y} \right) \frac{\partial v}{\partial x} + \sigma_x \frac{\partial}{\partial x} \left(\frac{\partial u}{\partial x} \right) \right.
$$

(h)

$$
\left. + \tau_{xy} \frac{\partial}{\partial x} \left(\frac{\partial u}{\partial y} + \frac{\partial v}{\partial x} \right) + \sigma_y \frac{\partial}{\partial x} \left(\frac{\partial v}{\partial y} \right) \right] dxdy
$$

No matter whether we have a linear elastic body or elastic–plastic body, the equilibrium condition is always established. Thus, the first two terms on the right-hand side of Eq. (h) are zero. By using the relation of strain and displacement, the remaining terms of Eq. (h) can be changed to

$$
\int_C T_i \frac{\partial u_i}{\partial x} ds = \int_A \left(\sigma_x \frac{\partial \varepsilon_x}{\partial x} + \sigma_y \frac{\partial \varepsilon_y}{\partial x} + 2\tau_{xy} \frac{\partial \varepsilon_{xy}}{\partial x} \right) dxdy
$$

$$
= \int_A \sigma_{ij} \frac{\partial \varepsilon_{ij}}{\partial x} dxdy
$$

(i)

Since Eq. (e) is equal to Eq. (i), the establishment of Eq. (b) is proved. Because $dy = 0$ on routes C_3 and C_4, and stress free on the crack surface at the same time, $T_x = 0$ and $T_y = 0$ respectively, which do not make a contribution to the integration value. In this way, the proof of establishment of Eq. (b) is equivalent to proving the setup of Eq. (a). Thus, that J integration is independent of the route has been proven.

Because of the J integral's independence of the route, the values of the J integral for many fracture problems can be easily calculated by selecting a suitable integral route. Two examples are shown in Figure B.6, for an infinitely long plate containing a symmetrical semi-infinite crack: (a) fixed

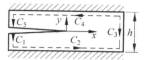

FIGURE B.6 **An infinitely long plate containing a symmetric semi-infinite crack.** (a) Fixed displacements on the boundaries $y = \pm h/2$. (b) Fixed moment of forces M acting on the free ends of upper and lower cantilever beams respectively.

displacements on the boundaries $y = \pm h/2$; (b) fixed moment of forces M acting on the free ends of the upper and lower cantilever beams respectively, which are separated by a crack.

For these two examples, the integral route is taken very close to the outer edge of a long plate and the configuration is an infinitely long rectangle, so the total route C can be divided into five segments, C_1 to C_5.

(1) Fixed displacements on the boundaries $y = \pm h/2$. Integration at points C_1 and C_5 respectively: it is a very far distance from the crack tip, so fixed displacement can only make segments C_1 and C_5 produce rigid movement. Therefore, the surface force on the segment is always zero, and the strain energy density is also zero and makes no contribution to the integral result.

Integration at points C_2 and C_4 respectively: $dy = 0$ and $u = v = $ constant, so there is no contribution to the integral result.

Integration at point C_3: $n_x = 1$, $n_y = 0$, $dy = ds$, $\sigma_x = 0$, and $\tau_{xy} = 0$, but $\sigma_y \neq 0$, so $T_x = T_y = 0$, $W \neq 0$.

$$J = \int_C W dy - T_i \frac{\partial u_i}{\partial x} ds = \int_{C_3} W dy - T_i \frac{\partial u_i}{\partial x} ds = \int_{-h/2}^{h/2} W dy = W_\infty h$$

Since point C_3 is located at an infinite distance, in the situation of fixed displacement as shown in Figure B.6(a), σ_y and ε_y are constant, so W is also constant. Here, W_∞ indicates the strain energy density of the constant value at this point.

(2) Fixed moment of forces M acting on the free ends of upper and lower cantilever beams. Integration at points C_2 and C_4: $dy = 0$, $T_x = T_y = 0$, so there is no contribution to the integral result.

Integration at point C_3: there is no effect of moment of force M, without stress and strain energy density, so it makes no contribution to the integral result.

Integration at points C_1 and C_5: $n_x = -1$, $n_y = 0$, $dy = -ds$, $\sigma_x \neq 0$, but $\tau_{xy} = \sigma_y = 0$. Now, we have

$$T_i \frac{\partial u_i}{\partial x} = -\sigma_x \varepsilon_x$$

thus

$$J = \int_C W\,dy - T_i\frac{\partial u_i}{\partial x}\,ds = \int_{C_1+C_5} W\,dy - T_i\frac{\partial u_i}{\partial x}\,ds$$

$$= \int_{h/2}^{-h/2} (W - \sigma_x\varepsilon_x)\,dy = \int_{-h/2}^{h/2} (\sigma_x\varepsilon_x - W)\,dy$$

Since $(\sigma_x\varepsilon_x - W)$ is the complementary energy density, by supposing $W^*_{-\infty}$ is the average complementary energy density at points C_1 and C_5 respectively (equivalently $x \to -\infty$), then it is obtained by

$$J = W^*_{-\infty}h$$

B.3 ENERGY EXPLANATION OF THE J INTEGRAL

In Chapter 2 and Appendix B.1 of this book, we described the physical significance of the J integral, which uses the same concept as Griffith's energy release rate; the J integral is a type of mechanical parameter based on the energy concept. In the deduction process of section B.1, the J integral of one side crack can be obtained under the situation of a given surface force, which is

$$J = -\frac{\partial}{\partial a}\left[\int_A W\,dA - \int_{C_1} T_i u_i\,ds\right] \tag{a}$$

Here, the arbitrary integral route C is replaced by a route C_1, which is very close to the plate edge. The whole plate area is A, which is surrounded by the route C_1, as shown in Figure B.7(a). This type of route substitution does not affect the value of the J integral. On the boundary Γ, a part of it is Γ_t, the surface force; another part of it is Γ_u, the displacement. Therefore, the linear integral part of Eq. (a) is only calculated on Γ_t, while the surface force on Γ_t is not affected by the quasi-static change of crack length. Thus, Eq. (a) becomes

$$J = -\frac{\partial}{\partial a}\left[\int_A W\,dA - \int_{\Gamma_t} T_i u_i\,ds\right] = \frac{\partial}{\partial a}[U - L] \tag{b}$$

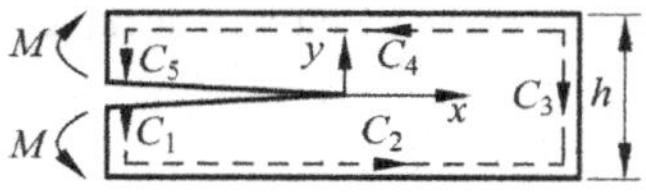

FIGURE B.7 A plate containing a crack. (a) A plate contains a side crack. (b) A plate contains an interior crack.

Here, U is the total strain energy of the plate and L is external work on the elastic–plastic plate, which are the integration values corresponding to the two terms in Eq. (a). According to the definition of total potential energy,

$$V = U - L \tag{B.7}$$

Thus, the J integral can be expressed as

$$J = -\frac{\partial V}{\partial a} \tag{B.8}$$

Equation (B.8) indicates that the J integral quantifies the potential energy variation in each unit thickness when the crack length changes by a unit length for an elastic–plastic plate containing the crack. The potential energy of the system always decreases as the crack increases, no matter what the boundary condition of the surface force or the displacement, such that the value of $\partial V/\partial a$ is negative. Consequently, the value of the J integral is always positive. When it obeys the linear elastic law, then $J = G$. Strictly speaking, the elastic–plastic plate in J integral application is a nonlinear elastic body or a simply loaded (without unloading) elastic–plastic body. If the unloading is applied during plastic deformation, the influence of loading history cannot be neglected. At present, Eq. (a) is not always established. In other words, Eq. (B.8) can only be used for an elastic body or a simply loaded elastic–plastic body.

If an elastic–plastic plate includes more than two interior crack tips, it is necessary to select an appropriate integral route along the positive direction (anticlockwise direction) surrounding each crack tip. However, the integral directions of two cracks are opposite at the interior of an elastic–plastic plate; as shown in Figure B.7(b), the overlapping routes result in counterbalance. Consequently, the sum of values of the J integral surrounding each crack tip is equivalent to the value of the J integral along an outer edge of the plate. At this time, Eq. (B.8) is still established, but the value represents the total value of the J integral of the system, rather than the integration value of a single crack tip.

Since Eq. (B.8) can be used by the J integral to express the energy significance, the variation of total potential energy following the change of crack length can be used to demarcate the relation between J integration and crack length for an elastic–plastic plate containing only one crack tip or having the same integral value at each crack tip. Consider a unit thickness rectangular plate with a single-side crack a, as shown in Figure B.8(a). The boundary condition is that one end is fixed and the other end is given load or displacement.

(1) Given load. As shown in Figure B.8(b), when $a \rightarrow a + \Delta a$, ΔV is negative and the absolute value is equal to an area OAB. By taking a small integral element of the marked area, the calculation is given by

$$[\delta(a + \Delta a) - \delta(a)]\mathrm{d}P$$

FIGURE B.8 Demarcation of J integration. (a) Sample. (b) Given load.

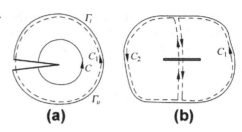

(a)   **(b)**

From this, the integral is obtained as

$$\Delta V = - \int_{O}^{P} [\delta(a + \Delta a) - \delta(a)] dP = - \int_{O}^{P} \left(\frac{\partial \delta}{\partial a}\right) \Delta a \, dP$$

The value of the J integral becomes

$$J = - \lim_{\Delta a \to 0} \frac{\Delta V}{\Delta a} = \int_{O}^{P} \left(\frac{\partial \delta}{\partial a}\right)_{P} dP \qquad (B.9)$$

(2) Given displacement. As shown in Figure B.8(c), when $a \to a + \Delta a$, ΔV is negative and the absolute value is equal to an area OAB. By taking a small integral element of the marked area, the calculation is given by

$$-[P(a + \Delta a) - P(a)] d\delta$$

From this, the integral is obtained as

$$\Delta V = \int_{O}^{\delta} [P(a + \Delta a) - P(a)] d\delta = \int_{O}^{\delta} \left(\frac{\partial P}{\partial a}\right)_{\delta} \Delta a \, d\delta$$

The value of the J integral becomes

$$J = - \lim_{\Delta a \to 0} \frac{\Delta V}{\Delta a} = - \int_{O}^{\delta} \left(\frac{\partial P}{\partial a}\right)_{\delta} d\delta \qquad (B.10)$$

B.4 CRACK INITIATION CRITERION OF THE J INTEGRAL

For a nonlinear elastic body containing a crack, the crack initiation criterion of J integration is an extension of the energy release rate criterion. For an elastic–plastic body containing the crack, we must be careful to use the J

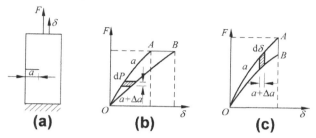

FIGURE B.9

integral criterion. There is a quite large plastic deformation frequently occurring at the crack tip location or an even larger scope in the materials of an elastic–plastic body before rupture. Consequently, the plastic deformation before crack initiation must be overcome or crack propagation leading to unstable rupture could occur. This kind of propagation after crack initiation usually implies a subcritical crack. Therefore, by following subcritical crack propagation, the local unloading unavoidably occurs at the crack tip location. At present, whether the *J* integral can be continually used as the mechanical parameter to measure the intensity of the stress and strain fields at crack tips warrants further study. So, the *J* integral criterion can only be used for the crack initiation criterion of a fractured elastic–plastic body under simple loading, but it cannot be used for the unstable rupture criterion. Readers should consult Chapter 2 for a discussion of unstable crack propagation.

For a nonlinear elastic body and elastic–plastic body under simple loading, the crack initiation criterion is

$$J \geq J_i \tag{B.11}$$

where J_i is the *J* integral value of crack initiation.

Since a mode I crack is the most common fracture, the theory of the *J* integral is mainly focused on mode I cracks. The *J* integral value of crack initiation is expressed by J_{Ic} for crack unstable propagation occurring soon after initiation. Thus, the *J* integral initiation criterion or fracture criterion is

$$J_I \geq J_{Ic} \tag{B.12}$$

It is strictly required for the sample dimensions that the constant J_{Ic} be measured for most metallic materials. The value of J_{Ic} for a brittle material is equivalent to the value of G_{Ic} for the plane strain problem. Following an increase of material toughness or lacking a standard of sample size, the value of J_i may deviate from the value of G_{Ic}. Therefore, $J_{Ic} = K_{Ic}^2/E_I$ is used to convert K_{Ic}, which is only established when brittle fracture occurs. Figure B.9

References

Abraham, F., Schneider, D., Land, B., Lifka, D., Skovira, J., Gerner, J., & Rosenkrantz, M. (1997). Instability dynamics in three-dimensional fracture: an atomistic simulation. *J. Mech. Phys. Solids, 45*, 1461.

Ahmad, S., Irons, B. B., & Zienkiewicz, O. C. (1970). Analysis of thick and thin shell structures by curved finite elements. *Int. J. Numer. Meth. Eng., 2*, 419–451.

Amirkhizi, A., et al. (2006). An experimentally-based viscoelastic constitutive model for polyurea, including pressure and temperature effects. *Phil. Mag., 86*(36), 5847–5866.

Areias, P., & Belytschko, T. (2005a). Nonlinear analysis of shells with arbitrary evolving cracks using XFEM. *Int. J. Numer. Meth. Eng., 62*(3), 384–415.

Areias, P., & Belytschko, T. (2005b). Analysis of three-dimensional crack initiation and propagation using the extended finite element method. *Int. J. Numer. Meth. Eng., 63*(5), 760–788.

Areias, P., & Belytschko, T. (2006). Two-scale shear band evolution by local partition of unity. *Int. J. Numer. Meth. Eng., 66*, 878–910.

Areias, P., & Belytschko, T. (2007). Two scale method for shear bands: thermal effects and variable bandwidth. *Int. J. Numer. Meth. Eng., 72*(6), 658–696.

Ashwell, D. G., & Sabir, A. B. (1972). New cylindrical-shell finite element based on simple independent strain functions. *Int. J. Mech. Sci., 14*, 171–183.

Attigui, M., & Petit, C. (1997). Mixed-mode separation in dynamic fracture mechanics: New path independent integrals. *Int. J. Fracture, 84*, 19–36.

Bachlechner, M., Omeltchenko, A., Nakano, A., Kalia, R., Vashishta, P., Ebbsjö, I., Madhukar, A., & Messina, P. (1998). Multimillion-atom molecular dynamics simulation of atomic level stresses in Si (111)/Si$_3$N$_4$ (0001) nanopixels. *Appl. Phys. Lett., 72*, 1969.

Bathe, K. J., & Ho, L. W. (1981). A simple and effective element for analysis of general shell structures. *Comput. Struct., 13*, 673–681.

Belytschko, T., & Black, T. (1999). Elastic crack growth in finite elements with minimal remeshing. *Int. J. Numer. Meth. Eng., 45*(5), 601–620.

Belytschko, T., & Tabbara, M. (1996). Dynamic fracture using element-free Galerkin methods. *Int. J. Numer. Meth. Eng., 39*(6), 923–938.

Belytschko, T., Fish, J., & Engelmann, B. E. (1988). A finite element with embedded localization zones. *Comput. Meth. Appl. Mech. Eng., 70*(1), 59–89.

Belytschko, T., Gu, L., & Lu, Y. (1994). Fracture and crack growth by element free Galerkin methods. *Modelling Simulation Mater. Sci. Eng., 2*, 519.

Belytschko, T., Liu, W., & Moran, B. (2000). *Nonlinear finite elements for continua and structures.* Wiley.

Belytschko, T., et al. (2001). Arbitrary discontinuities in finite elements. *Int. J. Numer. Meth. Eng., 50*(4), 993–1013.

Belytschko, T., Moës, N., & Gravouil, A. (2002). Non-planar 3D crack growth by the extended finite element and level sets—Part I: Mechanical model. *Int. J. Numer. Meth. Eng., 53*, 2549–2568.

Belytschko, T., et al. (2003a). Dynamic crack propagation based on loss of hyperbolicity and a new discontinuous enrichment. *Int. J. Numer. Meth. Eng., 58*, 1873–1905.

Belytschko, T., et al. (2003b). Structured extended finite element methods for solids defined by implicit surfaces. *Int. J. Numer. Meth. Eng.*, *56*(4), 609–635.

Belytschko, T., & Gracie, R. (2007). On XFEM applications to dislocations in problems with interfaces. *Int. J. Plasticity*, *23*, 1721–1738.

Belytschko, T., Gracie, R., & Ventura, G. (2009). A review of extended/generalized finite element methods for material modeling. *Modelling Simulation Mater. Sci. Eng.*, *17*, 043001.

Benvenuti, E., Tralli, A., & Ventura, G. (2008). A regularized XFEM model for the transition from continuous to discontinuous displacements. *Int. J. Numer. Meth. Eng.*, *74*(6), 911–944.

Biwa, S. (2001). Independent scattering and wave attenuation in viscoelastic composites. *Mech. Mater.*, *33*(11), 635–647.

Biwa, S., Idekoba, S., & Ohno, N. (2002). Wave attenuation in particulate polymer composites: independent scattering/absorption analysis and comparison to measurements. *Mech. Mater.*, *34*(10), 671–682.

Blandford, G. E., Ingraffea, A. R., & Liggett, J. A. (1981). Two dimensional stress intensity factor computations using the boundary element method. *Int. J. Numer. Meth. Eng.*, *17*(3), 387–404.

Böhme, W., & Kalthoff, J. (1982). The behavior of notched bend specimens in impact testing. *Int J Fract*, *20*, 139–143.

Borden, M. J., Verhoosel, C. V., Scott, M. A., Hughes, T. J. R., & Landis, C. M. (2012). A phase-field description of dynamic brittle fracture. *Comput. Meth. Appl. Mech. Eng.*, *217*, 77–95.

Buetchter, N., & Ramm, E. (1992). Shell theory versus degeneration — a comparison of large rotation finite element analysis. *Int. J. Numer. Meth. Eng.*, *34*, 39–59.

Cai, W., Arsenlis, A., Weinberger, C. R., & Bulatov, V. V. (2006). A non-singular continuum theory of dislocations. *J. Mech. Phys. Solids*, *54*, 561–587.

Cantin, G., & Clough, R. W. (1968). A curved cylindrical-shell finite element. *AIAA J.*, *6*, 1057.

Capdeville, Y., et al. (2003). Coupling the spectral element method with a modal solution for elastic wave propagation in global Earth models. *Geophys. J. Int.*, *152*(1), 35–67.

Chang, J., Xu, J., & Mutoh, Y. (2006). A general mixed-mode brittle fracture criterion for cracked materials. *Eng. Fracture Mech.*, *73*(9), 1249–1263.

Chessa, J., & Belytschko, T. (2003a). An extended finite element method for two-phase fluids: Flow simulation and modeling. *J. Appl. Mech.*, *70*(1), 10–17.

Chessa, J., & Belytschko, T. (2003b). An enriched finite element method and level sets for axisymmetric two-phase flow with surface tension. *Int. J. Numer. Meth. Eng.*, *58*(13), 2041–2064.

Chessa, J., & Belytschko, T. (2004). Arbitrary discontinuities in space–time finite elements by level sets and X-FEM. *Int. J. Numer. Meth. Eng.*, *61*(15), 2595–2614.

Chessa, J., Smolinski, P., & Belytschko, T. (2002). The extended finite element method (XFEM) for solidification problems. *Int. J. Numer. Meth. Eng.*, *53*(8), 1959–1977.

Chessa, J., Wang, H., & Belytschko, T. (2003). On the construction of blending elements for local partition of unity enriched finite elements. *Int. J. Numer. Meth. Eng.*, *57*(7), 1015–1038.

Chessa, J., & Belytschko, T. (2006). A local space-time discontinuous finite element method. *Comput. Method Appl. M.*, *195*, 1325–1343.

Clift, R., Grace, J., & Weber, M. (1978). Bubbles, drops, and particles. New York: Academic Press.

Colombo, D., & Giglio, M. (2006). A methodology for automatic crack propagation modelling in planar and shell FE models. *Eng. Fracture Mech.*, *73*, 490–504.

Coppola-Owen, A. H., & Codina, R. (2005). Improving Eulerian two-phase flow finite element approximation with discontinuous gradient pressure shape functions. *Int. J. Numer. Meth. Fluids*, *49*(12), 1287–1304.

Csikor, F., Motz, C., Weygand, D., Zaiser, M., & Zapperi, S. (2007). Dislocation avalanches, strain bursts, and the problem of plastic forming at the micrometer scale. *Science, 318,* 251.

Daux, C., Moës, N., Dolbow, J., Sukumar, N., & Belytschko, T. (2000). Arbitrary branched and intersecting cracks with extended finite element method. *Int. J. Numer. Meth. Eng., 48,* 1741–1760.

Duan, Q., et al. (2009). Element-local level set method for three-dimensional dynamic crack growth. *Int. J. Numer. Meth. Eng., 80*(12), 1520–1543.

England, A. H. (1965). A crack between dissimilar media. *ASME J. Appl. Mech., 32,* 400–402.

Erdogan, F. (1965). Stress distribution in bonded dissimilar materials with cracks. *ASME J. Appl. Mech., 32,* 403–410.

Farsad, M., Vernerey, F. J., & Park, H. S. (2010). An extended finite element/level set method to study surface effects on the mechanical behavior and properties of nanomaterials. *Int. J. Numer. Meth. Eng., 84*(12), 1466–1489.

Feder, J. (1980). Random sequential adsorption. *J. Theor. Biol., 87*(2), 237–254.

Finegan, I. C., & Gibson, R. F. (1999). Recent research on enhancement of damping in polymer composites. *Compos. Struct., 44*(2–3), 89–98.

Freund, L. B. (1990). *Dynamic fracture mechanics.* New York: Cambridge University Press.

Fries, T. P. (2008). A corrected XFEM approximation without problems in blending elements. *Int. J. Numer. Meth. Eng., 75*(5), 503–532.

Fries, T. P., & Belytschko, T. (2006). The intrinsic XFEM: a method for arbitrary discontinuities without additional unknowns. *Int. J. Numer. Meth. Eng., 68*(13), 1358–1385.

Gao, Y., Zhuang, Z., Liu, Z. L., You, X. C., Zhao, X. C., & Zhang, Z. H. (2011). Investigations of pipe-diffusion-based dislocation climb by discrete dislocation dynamics. *Int. J. Plasticity, 27,* 1055–1071.

Gracie, R., & Belytschko, T. (2009). Concurrently coupled atomistic and XFEM models for dislocations and cracks. *Int. J. Numer. Meth. Eng., 78*(3), 354–378.

Gracie, R., Oswald, J., & Belytschko, T. (2008). On a new extended finite element method for dislocations: Core enrichment and nonlinear formulation. *J. Mech. Phys. Solids, 56*(1), 200–214.

Griffith, A. A. (1920). The phenomena of flow and rupture in solids. *Phil. Trans. R. Soc. (London), A221,* 163–198.

Hwang, K. Z., Xia, Z. X., Xue, M. D., & Ren, W. M. (1988). *Plate and shell theory.* Beijing: Tsinghua University Press (in Chinese).

Hu, K., Chandra, A., & Huang, Y. (1993). Multiple void-crack interaction. *Int. J. Solids Structures, 30*(11), 1473–1489.

Hughes, T. J. R., & Liu, W. K. (1981a). Nonlinear finite element analysis of shells: Part 1. Two-dimensional shells. *Comput. Meth. Appl. Mech. Eng., 26,* 167–181.

Hughes, T. J. R., & Liu, W. K. (1981b). Nonlinear finite element analysis of shells: Part 2. Three-dimensional shells. *Comput. Meth. Appl. Mech. Eng., 26,* 331–362.

Hui, C. Y., & Zehnder, A. T. (1993). A theory for the fracture of thin plates subjected to bending and twisting moments. *Int. J. Fract, 61,* 211–229.

Hussain, M., Pu, S., & Underwood, J. (1974). Strain energy release rate for a crack under combined Mode I and Mode II. *Fracture Analysis, 1974,* 1.

Hutchinson, J. W., Mear, M., & Rice, J. R. (1987). Crack paralleling an interface between dissimilar materials. *ASME J. Appl. Mech., 54,* 828–832.

Irwin, G. R. (1957). Analysis of stresses and strains near the end of a crack traveling a plate. *J. Appl. Mech., 24,* 361–364.

Irwin, G. R. (1969). Basic concepts for dynamic fracture testing. *J. Basic Eng., 91*, 519−524.

Irwin, G. R., & Corten, H. T. (1968). Evaluating the feasibility of basing pipeline operating pressure on in-place hydrostatic test pressure. *Report to Northern Natural Gas Company and El Paso Natural Gas Company.*

John, R. (1990). Mixed mode fracture of concrete subjected to impact loading. *J. Struct. Eng., 116*, 585.

Kalthoff, J. (1985). On the measurement of dynamic fracture toughnesses review of recent work. *Int. J. Fracture, 27*(3), 277−298.

Kanninen, M. F., & Popelar, C. H. (1985). *Advanced fracture mechanics*. New York: Oxford University Press.

Karniadakis, G., & Sherwin, S. (1999). *Spectral/HP element methods for CFD*. New York: Oxford University Press.

Kim, J. (2010). Models for wave propagation in two-dimensional random composites: A comparative study. *J. Acoust. Soc. Am., 127*, 2201.

Kim, T. H., Park, J. G., Choi, J. H., et al. (2010). Nonlinear dynamic analysis of reinforced concrete shell structures. *Struct. Eng. Mech., 34*, 686−702.

Kinra, V. K., Petraitis, M. S., & Datta, S. K. (1980). Ultrasonic wave propagation in a random particulate composite. *Int. J. Solids Struct., 16*(4), 301−312.

Kubin, L., & Canova, G. (1992). The modelling of dislocation patterns. *Scripta Metall. Mater., 27*, 957.

Landau, L., Wrobel, L. C., & Ebecken, N. F. F. (1978). Elastic−plastic analysis of shell structures. *Comput. Struct., 9*, 351−358.

Lee, H., & Krishnaswamy, S. (2000). Quasi-static propagation of sub-interfacial cracks. *Trans. ASME J. Appl. Mech., 67*, 444−452.

Lee, Y., & Freund, L. (1990). Fracture initiation due to asymmetric impact loading of an edge cracked plate. *J. Appl. Mech., 57*, 105−111.

Legay, A., Wang, H., & Belytschko, T. (2005). Strong and weak arbitrary discontinuities in spectral finite elements. *Int. J. Numer. Meth. Eng., 64*(8), 991−1008.

Liu, C. T., & Jiang, C. P. (2000). *Plate and shell fracture mechanics*. Beijing: National Defence Industry Press (in Chinese).

Liu, X., Xiao, Q., & Karihaloo, B. (2004). XFEM for direct evaluation of mixed mode SIFs in homogeneous and bimaterials. *Int. J. Numer. Meth. Eng., 59*(8), 1103−1118.

Liu, Z. L., Liu, X. M., Zhuang, Z., & You, X. C. (2008). A mesoscale investigation of strain rate effect on dynamic deformation of single-crystal copper. *Int. J. Solids Struct., 45*, 3674.

Liu, Z. L., Liu, X. M., Zhuang, Z., & You, X. C. (2009a). Atypical three-stage-hardening mechanical behavior of Cu single-crystal micropillars. *Scripta Mater., 60*, 594.

Liu, Z. L., Liu, X. M., Zhuang, Z., & You, X. C. (2009b). A multi-scale computational model of crystal plasticity at submicron-to-nanometer scales. *Int. J. Plasticity, 25*, 1436.

Liu, Z. L., Menouillard, T., & Belytschko, T. (2011). An XFEM/spectral element method for dynamic crack propagation. *Int. J. Fracture, 169*(2), 183−198.

Liu, Z. L., Oswald, J., & Belytschko, T. (2013). XFEM modeling of ultrasonic wave propagation in polymer matrix particulate/fiberous composites. *Wave Motion, 50*, 389−401.

Melenk, J. M., & Babuska, I. (1996). The partition of unity finite element method: basic theory and applications. *Comput. Meth. Appl. Mech. Eng., 139*(1−4), 289−314.

Menouillard, T., et al. (2006). Efficient explicit time stepping for the extended finite element method (X-FEM). *Int. J. Numer. Meth. Eng., 68*, 911−939.

Menouillard, T., Song, J., Duan, Q., & Belytschko, T. (2010). Time dependent crack tip enrichment for dynamic crack propagation. *Int J Fract, 162*, 33−49.

Miehe, C., & Gürses, E. (2007). A robust algorithm for configurational force driven brittle crack propagation with R adaptive mesh alignment. *Int. J. Numer. Meth. Eng., 72*(2), 127−155.

Mier-Torrecilla, M., Idelsohn, S. R., & Oñate, E. (2010). Advances in the simulation of multi-fluid flows with the particle finite element method: Application to bubble dynamics. *Int. J. Numer. Meth. Fluids, 67.*

Minev, P. D., Chen, T., & Nandakumar, K. (2003). A finite element technique for multifluid incompressible flow using Eulerian grids. *J. Comput. Phys., 187*(1), 255−273.

Moës, N., Dolbow, J., & Belytschko, T. (1999). A finite element method for crack growth without remeshing. *Int. J. Numer. Meth. Eng., 46*(1), 131−150.

Moës, N., et al. (2003). A computational approach to handle complex microstructure geometries. *Comput. Meth. Appl. Mech. Eng., 192*(28−30), 3163−3177.

Mohan, R. (1998). Fracture analyses of surface-cracked pipes and elbows using the line-spring/shell model. *Eng. Fracture Mech., 59*, 426−438.

Mott, N. F. (1948). Fracture of metals: Theoretical consideration. *Engineering, 165*, 16−18.

Moseley, P., Oswald, J., & Belytschko, T. (2012). Adaptive atomistic-to-continuum modeling of propagating defects. *Int. J. Numer. Meth. Eng., 92*, 835−856.

Needleman, A. (1999). Computational mechanics at the mesoscale. *Acta Mater., 48*, 105.

Neuhauser, H. (1983). Slip-line formation and collective dislocation motion. *Dislocations in Solids, 6*, 319.

Nikishkov, G., & Atluri, S. (1987). Calculation of fracture mechanics parameters for an arbitrary three dimensional crack, by the 'equivalent domain integral' method. *Int. J. Numer. Meth. Eng., 24*(9), 1801−1821.

Nuismer, R. (1975). An energy release rate criterion for mixed mode fracture. *Int. J. Fracture, 11*(2), 246−250.

Oñate, E., Valls, A., & García, J. (2006). FIC/FEM formulation with matrix stabilizing terms for incompressible flows at low and high Reynolds numbers. *Comput. Mech., 38*(4), 440−455.

Palaniswamy, K., & Knauss, W. (1978). On the problem of crack extension in brittle solids under general loading. *Mech. Today, 4*, 87−148.

Pan, Y., Iorga, L., & Pelegri, A. A. (2008). Numerical generation of a random chopped fiber composite RVE and its elastic properties. *Compos. Sci. Technol., 68*(13), 2792−2798.

Parisch, H. (1995). A continuum-based shell theory for non-linear applications. *Int. J. Numer. Meth. Eng., 38*, 1855−1883.

Patera, A. (1984). A spectral element method for fluid dynamics: laminar flow in a channel expansion. *J. Comput. Phys., 54*(3), 468−488.

Réthoré, J., Gravouil, A., & Combescure, A. (2005). A combined space−time extended finite element method. *Int. J. Numer. Meth. Eng., 64*(2), 260−284.

Reusken, A. (2008). Analysis of an extended pressure finite element space for two-phase incompressible flows. *Comput. Visual. Sci., 11*(4), 293−305.

Ribeaucourt, R., Baietto-Dubourg, M. C., & Gravouil, A. (2007). A new fatigue frictional contact crack propagation model with the coupled X-FEM/LATIN method. *Comput. Meth. Appl. Mech. Eng., 196*(33−34), 3230−3247.

Rice, J. R. (1988). Elastic Fracture Mechanics Concepts for Interfacial Cracks. *ASME J. Appl. Mech., 55*, 98−103.

Rozycki, P., et al. (2008). X-FEM explicit dynamics for constant strain elements to alleviate mesh constraints on internal or external boundaries. *Comput. Meth. Appl. Mech. Eng., 197*(5), 349−363.

Samaniego, E., & Belytschko, T. (2005). Continuum−discontinuum modelling of shear bands. *Int. J. Numer. Meth. Eng., 62*(13), 1857−1872.

Sauerland, H., & Fries, T.-P. (2011). The extended finite element method for two-phase and free-surface flows: A systematic study. *J. Comput. Phys., 230*(9), 3369−3390.

Schladitz, K., et al. (2006). Design of acoustic trim based on geometric modeling and flow simulation for non-woven. *Comput. Mater. Sci., 38*(1), 56−66.

Segurado, J., & Llorca, J. (2002). A numerical approximation to the elastic properties of sphere-reinforced composites. *J. Mech. Phys. Solids, 50*(10), 2107−2121.

Seriani, G., & Oliveira, S. (2007). Dispersion analysis of spectral element methods for elastic wave propagation. *Wave Motion, 45*(6), 729−745.

Shakib, F., Hughes, T. J. R., & Johan, Z. (1991). A new finite element formulation for computational fluid dynamics: X. The compressible Euler and Navier−Stokes equations. *Comput. Meth. Appl. Mech. Eng., 89*(1−3), 141−219.

Shivakumar, K., & Raju, I. (1992). An equivalent domain integral method for three-dimensional mixed-mode fracture problems. *Eng. Fracture Mech., 42*(6), 935−959.

Simo, J. C., & Fox, D. D. (1989). On a stress resultant geometrically exact shell model, Part I: Formulation and optimal parameterization. *Comput. Meth. Appl. Mech. Eng., 72*, 267−304.

Smolianski, A. (2001). *Numerical modeling of two-fluid interfacial flows. Dissertation.* University of Jyväskylä.

Smolianski, A. (2005). Finite-element/level-set/operator-splitting (FELSOS) approach for computing two-fluid unsteady flows with free moving interfaces. *Int. J. Numer. Meth. Fluids, 48*(3), 231−269.

Song, J., & Belytschko, T. (2009). Dynamic fracture of shells subjected to impulsive loads. *J. Appl. Mech., 76*, 051301.

Song, J., Areias, P., & Belytschko, T. (2006). A method for dynamic crack and shear band propagation with phantom nodes. *Int. J. Numer. Meth. Eng., 67*, 863−893.

Stolarska, M., et al. (2001). Modelling crack growth by level sets in the extended finite element method. *Int. J. Numer. Meth. Eng., 51*(8), 943−960.

Strouboulis, T., Copps, K., & Babu_ska, I. (2001). The generalized finite element method: an example of its implementation and illustration of its performance. *International Journal for Numerical Methods in Engineering, 47*(8), 1401−1417.

Sukumar, N., et al. (2000). Extended finite element method for three dimensional crack modelling. *Int. J. Numer. Meth. Eng., 48*(11), 1549−1570.

Sukumar, N., et al. (2001). Modeling holes and inclusions by level sets in the extended finite-element method. *Comput. Meth. Appl. Mech. Eng., 190*(46−47), 6183−6200.

Tezduyar, T. E., & Osawa, Y. (2000). Finite element stabilization parameters computed from element matrices and vectors. *Comput. Meth. Appl. Mech. Eng., 190*(3−4), 411−430.

Timoshenko, S., & Woinowsky-Krieger, S. (1959). *Theory of plates and shells* (2nd ed.). McGraw-Hill.

Tornberg, A., & Engquist, B. (2000). A finite element based level-set method for multiphase flow applications. *Comput. Visual. Sci., 3*(1), 93−101.

Tran, A., et al. (2011). A multiple level set approach to prevent numerical artefacts in complex microstructures with nearby inclusions within XFEM. *Int. J. Numer. Meth. Eng., 85*(11), 1436−1459.

Uchic, M., Dimiduk, D., Florando, J., & Nix, W. (2004). Sample dimensions influence strength and crystal plasticity. *Am. Assoc. Advancement Sci., 305*, 986.

Viz, M. J., Zehnder, A. T., & Bamford, J. D. (1995). Fatigue fracture of thin plates under tensile and transverse shear stresses. *Fracture Mechanics, 26*, 631−651.

Venkatesha, K. S., Ramamurthy, T. S., & Dattaguru, B. (1998). A study of the behaviour of sub-interfacial cracks in bi-material plates. *Eng. Fracture Mech., 59*, 241−252.

Ventura, G., Moran, B., & Belytschko, T. (2005). Dislocations by partition of unity. *Int. J. Numer. Meth. Eng., 62*(11), 1463–1487.

Wright, T. (2002). *The physics and mathematics of adiabatic shear bands*. Cambridge University Press.

Wu, C. H. (1978). Fracture under combined loads by maximum-energy-release-rate criterion. *J. Appl. Mech., 45*, 553.

Xiujun, F., Feng, J., & Jinting, W. (2007). Cohesive crack model based on extended finite element method. *Journal of Tsinghua University (Science and Technology), 47*(3), 344–347. in Chinese.

Xu, D. D., Zeng, Q. L., Liu, Z. L., & Zhuang, Z. (2013). Modeling of dynamic crack branching with extended finite element method. *Comput. Mech., submitted.*

Xu, X. P., & Needleman, A. (1994). Numerical simulations of fast crack growth in brittle solids. *J. Mech. Phys. Solids, 42*(9), 1397–1434.

Yang, M., & Kim, K. (1992). The behavior of subinterface cracks with crack-face contact. *Eng. Fracture Mech., 44*, 155–165.

Yoffe, E. H. (1951). The moving Griffth crack. *Phil. Mag., 42*, 739–750.

You, X. C., Zhuang, Z., Huo, C. Y., Feng, Y. R., & Zhuang, C. J. (2003). Crack arrest in a rupturing steel gas pipelines. *Int. J. Fracture, 123*(1–2), 1–14.

Zbib, H., & Diaz de la Rubia, T. (2002). A multiscale model of plasticity. *Int. J. Plasticity, 18*, 1133.

Zhou, S., Lomdahl, P., Voter, A., & Holian, B. (1998). Three-dimensional fracture via large-scale molecular dynamics. *Eng. Fracture Mech., 61*, 173.

Zhuang, Z. (1995). *The development of finite element methods for the investigation of dynamic crack propagation in gas pipelines. Ph.D. thesis.* Ireland: University College Dublin.

Zhuang, Z., & Cheng, B. B. (2011a). Development of X-FEM methodology and study on mixed-mode crack propagation. *Acta Mech. Sin., 27*(3), 406–615.

Zhuang, Z., & Cheng, B. B. (2011b). Equilibrium state of mode-I sub-interfacial crack growth in bi-materials. *Int. J. Fracture, 170*(1), 27–36.

Zhuang, Z., & Cheng, B. B. (2011c). A novel enriched CB shell element method for simulating arbitrary crack growth in pipes. *Science China Phys. Mech. Astron., 54*(8), 1520–1531.

Zhuang, Z., & O'Donoghue, P. E. (2000a). The recent development of analysis methodology for crack propagation and arrest in the gas pipelines. *Int. J. Fracture, 101*(3), 269–290.

Zhuang, Z., & O'Donoghue, P. E. (2000b). Determination of material fracture toughness by a computational/experimental approach for rapid crack propagation in PE pipe. *Int. J. Fracture, 101*(3), 251–268.

Zlotnik, S., & Diez, P. (2009). Hierarchical X-FEM for *n*-phase flow (*n* > 2). *Comput. Meth. Appl. Mech. Eng., 198*(30–32), 2329–2338.

Zucchini, A., Hui, C., & Zehnder, A. T. (2000). Crack tip stress fields for thin, cracked plates in bending, shear and twisting: A comparison of plate theory and three-dimensional elasticity theory solutions. *Int. J. Fracture, 104*(4), 387–407.

Further Reading

Areias, P., Song, J., & Belytschko, T. (2006). Analysis of fracture in thin shells by overlapping paired elements. *Comput. Meth. Appl. Mech. Eng., 195*(41–43), 5343–5360.

Belytschko, T., & Chessa, J. (2006). A local space–time discontinuous finite element method. *Comput. Meth. Appl. Mech. Eng., 195*(13–16), 1325–1343.

Cao, H. C., & Evans, A. G. (1989). An experimental study of the fracture resistance of bimaterial interfaces. *Mech. Mater., 7*, 295–304.

Charalambides, P. G., Lund, J., Evans, A. G., & McMeeking, R. M. A. (1989). Test specimen for determining the fracture resistance of a bimaterial interface. *J. Appl. Mech., 56*, 77–82.

Dundurs, J. (1969). Discussion of edge-bonded dissimilar orthogonal elastic wedges under normal and shear loading. *ASME J. Appl. Mech., 36*, 650–652.

Dundurs, J., & Mura, T. (1964). Interaction between an edge dislocation and a circular inclusion. *J. Mech. Phys. Solids, 12*, 177–189.

Dundurs, J., & Sendeckyi, G. P. (1965). Behavior of an edge dislocation near a bimetallic interface. *ASME J. Appl. Mech., 36*, 3353–3354.

Erdogan, F. (1971). Bonded dissimilar materials containing cracks parallel to the interface. *Eng. Fracture Mech., 3*, 231–240.

Fan, T. Y. (1990). *Introduction of dynamic fracture mechanics*. Beijing: Beijing Institute of Technology Press (in Chinese).

Foltyn, P. A., & Ravi-Chandar, K. (1993). Initiation of an interface crack under mixed-mode loading. *ASME J. Appl. Mech., 60*, 227–229.

Gato, C. (2010). Detonation-driven fracture in thin shell structures: Numerical studies. *Appl. Math. Model, 34*, 3741–3753.

Gracie, R., Ventura, G., & Belytschko, T. (2007). A new fast method for dislocations based on interior discontinuities. *Int. J. Numer. Meth. Eng., 69*, 423–441.

Gravouil, A., Moes, N., & Belytschko, T. (2002). Non-planar 3D crack growth by the extended finite element and level sets – Part II: Level set update. *Int. J. Numer. Meth. Eng., 53*, 2569–2586.

Hallquist, J. O. (1991). *LS-DYNA3D theoretical manual*. Livermore Software.

Hwang, K. Z., & Yu, S. W. (1985). *Elastic–plastic fracture mechanics*. Beijing: Tsinghua University Press (in Chinese).

Liechti, K. M., & Chai, Y. S. (1991). Biaxial loading experiments for determining interfacial fracture toughness. *ASME J. Appl. Mech., 58*, 680–687.

Liu, W. K. (2007). Simulation-based engineering and science approach to analysis and design of microsystems: From a dream to a vision to reality. In *Proceedings of the WTEC Workshop on U.S. R&D in simulation-based engineering and science* (pp. 77–95).

Lu, Y. Z. (1987). *Engineering fracture mechanics*. Xian: Xian Jiaotong University Press (in Chinese).

Mason, J. J., Lambros, J., & Rosakis, A. J. (1992). The use of a coherent gradient sensor in dynamic mixed-mode fracture mechanics experiments. *J. Mech. Phys. Solids, 40*, 641–661.

Moës, N., & Belytschko, T. (2002). Extended finite element method for cohesive crack growth. *Eng. Fracture Mech., 69*, 813–833.

Moës, N., Gravouil, A., & Belytschko, T. (2002). Non-planar 3D crack growth by the extended finite element and level sets – Part I: Mechanical model. *Int. J. Numer. Meth. Eng., 53*, 2549–2568.

Nishioka, T. (1997). Computational dynamic fracture mechanics. *Int. J. Fracture, 86*, 127–159.

Oden, J. T. (2007). The NSF Blue Ribbon Panel Report on SBES. In *Proceedings of the WTEC Workshop on U.S. R&D in simulation-based engineering and science* (pp. 1–11).

O'Dowd, N. P., Shih, C. F., & Stout, M. G. (1992). Test geometries for measuring interfacial fracture toughness. *Int. J. Solids Struct., 29*, 571–589.

O'Donoghue, P. E., & Zhuang, Z. (1999). A finite element model for crack arrestor design in gas pipelines. *Fatigue Fracture Eng. Mater. Struct., 22*(1), 59–66.

O'Donoghue, P. E., Green, S. T., Kanninen, M. F., & Bowles, P. K. (1991). The development of fluid/structure interaction model for flawed fluid containment boundaries with applications to gas transmission and distribution pipings. *Comput. Struct., 38*(5/6), 501–513.

Rabczuk, T., Areias, P., & Belytschko, T. (2007). A mesh-free thin shell method for non-linear dynamic fracture. *Int. J. Numer. Meth. Eng., 72,* 524–548.

Rice, J. R., Paris, P. C., & Merkle, J. G. (1973). Some further results of J-integral analysis and estimates. ASTM STP 536, Philadephia *Am. Soc. Testing and Materials,* 231–245.

Shih, C. F., Asaro, R., & O'Dowd, N. P. (1991). Elastic–plastic analysis cracks on bimaterial interfaces, Part III: Large-scale yielding. *ASME J. Appl. Mech., 58,* 450–463.

Suo, Z., & Hutchinson, J. W. (1989). Steady-state cracking in brittle substrates beneath adherent films. *Int. J. Solids Struct., 25,* 1337–1353.

Tippur, H. V., & Rosakis, A. J. (1991). Quasi-static and dynamic crack growth along bimaterial interfaces: A note on crack-tip field measurements using coherent gradient sensing. *Exp. Mech., 31,* 243–251.

Tvergaard, V., & Hutchinson, J. W. (1993). The influence of plasticity on mixed mode interface toughness. *J. Mech. Phys. Solids, 41,* 1119–1135.

Wang, J. S., & Suo, Z. (1990). Experimental determination of interfacial toughness curves using Brazil-nut-sandwiches. *Acta Metall. Mater., 38,* 1279–1290.

Wyart, E., Coulon, D., Duflot, M., Pardoen, P., Remacle, J.-F., & Lani, F. (2007). A substructured FE-shell/XFE-3D method for crack analysis in thin-walled structures. *Int. J. Numer. Meth. Eng., 72,* 757–779.

Xu, L., & Tippur, H. V. (1995). Fracture parameters for interfacial cracks: An experimental-finite element study of crack tip fields and crack initiation toughness. *Int. J. Fracture, 71,* 345–363.

Yao, X. F., Yeh, H. Y., & Xu, W. (2005). Dynamic initiation and propagation behaviour of sub-interfacial cracks in PMMA/aluminium bi-material system. *Fatigue Fracture Eng. Mater. Struct., 28,* 1191–1198.

Yang, W. (1995). *Macro-micro-scale fracture mechanics.* Beijing: National Defence Industry Press (in Chinese).

Zhang, X. (1992). *Solution methods of stress intensity factor on fracture mechanics.* Beijing: National Defence Industry Press (in Chinese).

Zhuang, Z., & Guo, Y. J. (1999). Analysis of dynamic fracture mechanism in gas pipelines. *Eng. Fracture Mech., 64,* 271–289.

Zhuang, Z., & Jiang, C. P. (2004). *Engineering fracture and damage.* Beijing: Mechanical Industry Press.

Zhuang, Z., & O'Donoghue, P. E. (1997). Material fracture toughness determination for polyethylene pipe materials using small scale test results. *Acta Mech. Sin., 13*(1), 63–80.

Zhuang, S., & Ravichandrana, G. (2003). An experimental investigation of shock wave propagation in periodically layered composites. *J. Mech. Phys. Solids, 51,* 245–265.

Zhuang, Z., Liu, Z. L., Cheng, B. B., & Liao, J. H. (2012). *Extended finite element method.* Tsinghua University Press.

Index

Note: Page Numbers followed by f indicate figures; t, tables; b, boxes.

A

Absolute value function, 58—60, 60f, 101, 170
Adaptive mesh, 2
Additional degree of freedom, 7, 9, 172—173, 226—227
Angular distribution function, 146
Attenuation coefficient, 181—185, 187

B

Belytschko—Lin—Tsay shell element, 10—11, 115—116, 116f
Bimaterial, 9, 11—12, 79, 83—84, 143—144, 149—151, 150f, 153—155, 159—160, 162—166
Blending element, 8, 58—60, 71, 172, 197
Boundary element method, 2—3, 5—7

C

Coarse grain method, 223—224
Cohesive crack, 9
Complex moduli, 181, 184
Consistency, 61, 63, 79—80, 86—87, 109, 144—145, 170, 198
Consistent mass matrix, 72
Constitutive law, 10—11, 118
Continuum-based (CB) shell, 11—12, 104, 115—120, 117f, 123, 125, 128—129, 136, 140—142
Continuum mechanics, 213—216
Crack arrest, 33—34, 38—39, 46—47
Crack branch, 3—4, 6f, 8, 10, 93—101, 99f—100f
Crack propagation, 1—2, 7—8, 10—15, 19, 21, 23, 26—27, 33—50, 34f, 41t, 46f—47f, 49f, 75—79, 83—84, 89, 91—93, 95—97, 101, 129—133, 135f, 137, 139—140, 139f, 141f, 142, 144, 153—154, 156—157, 159—162, 165—166, 169, 215—216, 223—224, 225f, 245, 250, 256—257
Curvature, 114—115, 190—191, 205—206

D

Deformation band, 224—227
Deformation localization, 224—228
Discrete dislocation plasticity, 215—219, 222—223
Discrete equation, 173
Dislocation, 9, 148—149, 151—152, 165, 213—226, 214f—216f, 218f, 221f—224f
Dislocation core, 4—5, 216—217, 220—223, 220f, 222f
Distance function, 54—56, 70, 94—95, 120—121, 170, 192—193
Divergence-free condition, 189—190
Dynamic fracture mechanics, 14—15, 33—35, 38—39, 45—50
Dynamic stress intensity factor, 36—38, 49—50, 89
Dynamic viscosity, 189—190, 194—195

E

Element shape function, 85f
Energy release rate, 13—17, 19, 21, 24, 26—28, 31, 35, 37, 49—50, 130—132, 134—135, 146
Enrichment shape function, 4—5, 8—9, 11—12, 31, 52—53, 55—60, 70, 71f, 72—73, 119, 121—123, 122f—123f, 125—128, 142, 195—202
Euler angle, 175—176
Explicit method, 192
Extended finite element method (X-FEM), 2—12, 4f, 6f, 10f, 14—15, 29—31, 38, 42, 49—53, 55—58, 60, 62—64, 67—73, 75—92, 76f, 80f, 95, 97—98, 101, 104, 114—115, 119—129, 136, 140—142, 144, 153—157, 165, 169, 172, 175, 177—179, 178f, 181—182, 187, 189, 194—200, 205, 211, 215—216, 219—223, 223f—224f, 226—228

F

Fiber, 10—11, 14, 116—117, 119—120, 129, 141—142, 168, 175—176, 179—180, 180f—181f, 184—187, 185f
Finite element discretization, 126, 200

Finite element method, 2–9, 11–12, 17–18, 41–42, 49–50, 53, 62–64, 67–69, 69f, 71, 79–80, 80f, 84, 104, 113–114, 115f, 125, 129, 140–141, 144, 154–157, 165, 168–169, 177–179, 178f–179f, 195–196, 198–199, 215–218, 223, 223f, 226, 228

Fracture criterion, 14–15, 19–29, 31, 75–76, 257

Fracture mechanics, 1–2, 9, 11–22, 24–25, 29, 31, 56–57, 93, 111, 245, 248, 250

Fracture toughness, 13–15, 19–21, 20t, 23, 26–27, 31, 35, 39–40, 42, 44–47, 49, 89, 153, 245

G

Generation mode, 39

Geometric stiffness, 128

H

Heaviside function, 8–9, 55–56, 57f, 58, 70–71, 71f, 193, 198, 220–222, 222f, 226–227

Hereditary integral, 176

Heterogeneous material, 4, 9, 169, 173

I

Immiscible flow, 11–12, 189, 194, 198, 211

Incomplete term, 198–199

Incompressible flow, 11–12, 189–190, 195, 197–198, 207, 211

Initial configuration, 60–62, 61f, 69f

Interaction integral, 11–12, 14–15, 29–31, 29f, 37, 38f, 75–76, 99

Interface capturing, 192

Interface crack, 143–144

Interface tracking, 192, 198

Isothermal flow, 197

J

J integral, 11–12, 130–132, 222, 245–257, 256f

K

Kirchhoff–Love shell, 10–11

L

Lamina coordinates, 119

Laplace–Young equation, 198

Level set method, 53–55, 54f, 58–59, 72–73, 169–175, 171f, 177, 181, 192–196, 193f, 200, 204–205, 219–220, 221f

L^2 projection, 200, 205–206

Lumped mass matrix, 67–68, 72–73, 86–87, 173–175

M

Master node, 117, 119–120, 128–129, 142

Maximum circumferential stress criterion, 96–97, 100

Maximum energy release rate criterion, 27–29, 31, 133–136, 142

Micro-scale crystal plasticity, 213–223

Mindlin–Reissner assumption, 116, 119–120

Mindlin–Reissner shell, 104–105, 140–141

Molecular dynamics (MD) simulation, 175, 213–215

Multi-scale simulation, 9, 11–12, 223–224, 228

N

Navier–Stokes (N-S) equations, 192, 194, 199–202, 206–207, 211

Newmark β scheme, 127

Newtonian flow, 189–190

Nodal external force, 67, 72, 127, 173

Nodal force, 63–64, 129

Node force release method, 2, 11–12, 41–43, 42f, 49–50

P

Partition of unity, 8, 51–53, 58, 71–73, 197

Peach–Koehler force, 222

Plane strain, 17, 19, 22–23, 25–28, 30, 87, 132, 134–135, 146–147, 229–230, 235, 248–250, 257

Plane stress, 2–3, 16–17, 19, 22–23, 25, 30, 116, 118, 132, 134–135, 230, 235, 250

Polymer matrix composites (PMCs), 11–12, 167–169, 168f, 187

Pressure pipe fracture, 1, 10

Pressure Stabilizing Petrov Galerkin (PSPG) method, 200, 202–204, 211

Prony series, 177, 181–183

Propagation mode, 39

P-type refinement, 53, 84

R

Ramp functions, 58

Random sequential adsorption (RSA) algorithm, 175, 182–185, 187

Rayleigh wave speed, 35—36, 89, 99—100
Regularized Heaviside function, 226—228
Resistance curve, 40
Ridge enrichment shape function,
    197—199
Row summation method, 173

S

Shear band, 9, 226—227, 226f
Shear instability, 226
Shell and plate fracture mechanics, 104—113,
    140—141
Shifted enrichment function, 58, 75—76
Signed distance, 54—56, 70, 94—95,
    120—121, 170, 192—193, 227f
Slave node, 117, 119—120, 128—129, 142
Spectral element, 9, 75—76, 84—89, 85f—86f,
    91—92, 92f, 101
Steady-state propagation, 44
Strain energy density, 16—17, 20—21, 23—26,
    31, 43—44, 130, 247—248, 251, 253
Streamline Upwind Petrov Galerkin (SUPG)
    method, 200, 202—204, 211
Stress intensity factor (SIF), 4—5, 7f, 9—12,
    14—28, 29, 18t, 30—31, 35—38, 42, 75—77,
    76f, 81, 87—89, 88f, 91—93, 91f—92f,
    98—99, 99t, 101, 104, 107—108, 111—112,
    115, 119, 129—136, 137f, 142—147, 149,
    152—153, 155—156, 158, 161—162,
    165—166, 229, 238, 240, 242—250

Stress singularity, 3—4, 17, 112—113,
    137—138, 146, 237
Stress wave oscillation, 91—92
Stress wave scattering, 183—185
Strong discontinuity, 4, 55—58, 60—61, 63,
    94, 194—195, 198, 216
STX-FEM, (Not Found)
Subdomain integration, 8—9, 64—68,
    65f—66f, 172
Subinterface crack, 9
Surface tension, 190—191, 194—195, 198,
    205, 208—209

T

Tip enrichment function, 10, 56—58, 57f
Total Lagrangian formulation, 125—127
Total stress, 189—191
Transient flow, 11—12, 189, 192, 198, 211
Two-phase flow, 11—12, 189, 191—200,
    203—204, 211

U

Updated Lagrangian formulation,
    (Not Found)

W

Weak discontinuity, 4, 11—12, 55, 58—61, 63,
    79, 168—169, 172, 194—200, 211
Weak forms, 60—64, 72—73, 125, 173